SHARON E.T PHILLIPS
19 LAKE DRIVE
LAKE IN THE HILLS, IL
60102-1352
PHONE. (708) 658-7042

Microwave
MESFETs and HEMTs

For a complete listing of the *Artech House Microwave Library*,
turn to the back of this book . . .

Microwave
MESFETs and HEMTs

J. Michael Golio
Editor
Motorola
Tempe, Arizona

Artech House
Boston • London

International Standard Book Number: 0-89006-426-1
Library of Congress Catalog Card Number: 91-4554

10   9   8   7   6   5   4   3   2   1

CONTENTS

PREFACE

It is the authors' conviction that book prefaces are read only by university students in the early hours of the morning. This exercise is typically undertaken immediately preceding an exam. The futile venture is prompted by a misguided and desperate hope that some knowledge—not attainable from multiple scourings of class notes and the course text—has perhaps been secretly hidden in an easy-to-understand formulation within the preface of the book. The reader, of course, fails to discover ultimate enlightenment in the preface, and is probably bored in the process. Like others before it, this preface offers little true illumination and is of minimal entertainment value. It does, however, briefly describe some of the topics covered in the book.

Most of the material presented in this book was developed or accumulated to improve the accuracy of microwave circuit performance predictions and to reduce the cycle time required for the design process. The authors worked together on these problems over a period of several years. Much of this effort was supported by internally funded research and development programs at the Motorola Government Electronics Group. Many of the ideas and most of the data presented in the book have evolved from the numerous discussions and collaborations that took place in this environment.

Chapter 1 provides a brief discussion of topics in solid-state theory and device physics, which are important to the operation of MESFETs and HEMTs. Minimal mathematical formulation is used in these explanations. Bulk semiconductor material properties are reviewed in Section 1.1. Subjects addressed include energy band diagrams and carrier transport. Both steady-state and transient transport topics are discussed. This is followed by material on semiconductor contacts. Ohmic contacts, p-n junctions and Schottky barriers are examined. The subject of heterojunctions is also approached in this section. Sections 1.3 and 1.5 are devoted to a basic description of the MESFET and HEMT, respectively. The defining physical structures and characteristic electrical properties are presented for both the MESFET and HEMT. Finally, some important second-order effects are discussed in Section 1.4. Low-frequency dispersion of the electrical characteristics, subthreshold effects and optical-device interactions are addressed. Although these effects are secondary to an understanding of MESFET and HEMT operation, many of these phenomena are central to the characterization and parameter extraction processes addressed in Chapters 3 and 4.

Device modeling is the subject of Chapter 2. The focus of this chapter is on MESFET and HEMT models that are incorporated or appropriate for incorporation into circuit simulation packages. Both empirically derived and physically based models are addressed. Section 2.1 reviews small-signal equivalent circuit modeling as well as a physically based MESFET model. Equivalent circuit model scaling issues are examined. The potential use of the physically based model for purposes of statistical design or device optimization is also explored. In Section 2.2, the subject of noise modeling is confronted. The behavior of important noise quantities is examined for both MESFETs and HEMTs. The predictions of various noise models is compared to direct measurements of noise performance. A physically derived noise model for the MESFET is used to illustrate the relationship between certain important physical properties of the device and the resulting overall noise performance. Finally, large-signal models are presented in Section 2.3. Several popular large-signal MESFET models are presented. A general technique for developing large-signal HEMT models from existing MESFET models is also described. This technique is applied to develop three different new HEMT models. Comparisons of the performance of all of the models are also presented. Physically based models are presented for both the MESFET and HEMT.

Chapter 3 focuses on device characterization. Characterization issues, as they relate to the parameter extraction and modeling process, are the important aspect of these discussions. A number of important dc measurements are presented in the first section of the chapter. Section 3.2 deals with RF measurement techniques as they might be applied for use with either small-signal or large-signal parameter extraction and modeling. The most important of these measurements is small-signal *S*-parameter characterization. Pulsed current-voltage and low-frequency dispersion measurement techniques are also described as part of this discussion. Direct measurements of large-signal device behavior are the topic of Section 3.3. Although these techniques present difficulties if applied to a model parameter extraction technique, the measurements can provide valuable information, which may be used for device performance evaluation. Both load-pull measurements and matched two-tone harmonic content measurements are described. In the last section of Chapter 3, noise characterization techniques are presented. Both direct measurement of minimum noise figure and abbreviated measurement techniques are discussed.

Algorithms for extracting model parameters from measured data are presented in Chapter 4. Extraction processes corresponding to most of the measurements described in Chapter 3 and for the models of Chapter 2 are described in detail. Comparisons of measured and modeled characteristics for both MESFETs and HEMTs are presented. Extraction of certain model parameters can be made very simply from measured dc data. These issues are the focus of Section 4.1. The techniques described for small-signal equivalent circuit extraction (Section 4.2) are significantly more accurate and efficient than traditional optimization methods. In Section 4.3, optimization techniques along with specific large-signal parameter extraction applications of those techniques are described. Some results from a particular large-signal parameter extraction routine described in this Section are also presented. A brief discussion of the fairly straightforward noise model extraction process concludes this chapter.

Section 5.1 presents an overview of microwave applications for MESFETs and HEMTs. This first section also presents a brief discussion of how the modeling techniques described in Chapters 2 through 4 are used in the design process for each application. Section 5.2 discusses device figures of merit (FOMs). Definitions and limitations in the use of these FOMs are also given along with information related to the utilization of device models to determine the figures of merit. The final two sections of Chapter 5 explore the historical performance obtained from both MESFETs and HEMTs, and we extrapolate this performance to the projected limits of the respective technologies.

ACKNOWLEDGEMENTS

The authors wish to thank Dave Halchin, Bill Rhyne, and Michael Dydyk for their comments and reviews of the manuscript. Thanks are also extended to Ray Waddoups, Bob Curlee, Warren Seely, and John Weidman for their support of this effort.

The editor also thanks his wife, JJ, for agreeing to rearrange, reschedule, or miss so many archaeological outings into the deserts of the Southwest while he worked on the manuscript. We both missed these diversions tremendously.

Chapter 1
BASIC OPERATION

The GaAs *metal-semiconductor field effect transistor* (MESFET) has been the "workhorse" of the microwave industry for many years. The MESFET is used as the active device for both low noise and power amplifiers as well as for transfer switches, attenuators, oscillators, and mixers. Fabrication of MESFETs is also compatible with the manufacture of monolithic circuitry. This fact ensures that interest in the MESFET will continue for some time. However, because the performance limits of the MESFET are being reached, other devices are being explored as substitutes for the MESFET. The *high electron mobility transistor* (HEMT) is one of the devices that offers improved performance as compared with MESFETs for many applications. Although the manufacturing technology for HEMTs is not as mature as that of MESFET fabrication at this time, the device has already exhibited performance not attainable from MESFETs and advances are being made rapidly.

The overall electrical behavior of microwave MESFETs and HEMTs is determined largely by the electrical properties of the bulk semiconductor material as well as the nature of the physical contacts to the material. For many years, these subjects have been primarily the concern of solid-state device engineers responsible for the design of microwave transistors. Increasingly, microwave circuit designers are also finding that a knowledge of solid-state physics is important to the successful completion of their assignments. The expanding role of *monolithic microwave integrated circuits* (MMICs) in the microwave industry is one of the forces driving this new interest.

This chapter presents a very brief discussion of some of the solid-state concepts that are involved in the accurate characterization, parameter extraction, and modeling of microwave MESFETs and HEMTs. We also examine the properties of terminal contacts and semiconductor heterostructures. Basic descriptions of device operation for both the MESFET and HEMT are addressed. The material presented in this chapter is important to an engineer interested in optimal utilization of the techniques presented in the remainder of the book. Although these techniques will be useful to both the microwave circuit designer and the device fabrication engineer, the material in this chapter is not intended to be a complete guide to those with an interest in optimizing device fabrication processes.

1.1 MATERIALS PROPERTIES

1.1.1 Energy Bands

The semiconductors of importance in the fabrication of microwave MESFETs and HEMTs are single crystal structures. Individual atoms of the material bond together to produce a structure that repeats periodically in any given direction. Models of the bonding in these crystalline solids are typically discussed in terms of energy-momentum relationships. By making assumptions regarding the material and the nature of the potential within the crystal, Schrodinger's equation can be solved and the energy-momentum (E-k) relationship [1] determined.

For semiconductor materials, the energy-momentum solution predicts a forbidden energy gap that contains no allowed states within which electrons can exist. Above the forbidden gap, electrons can exist in a number of allowed energy states. This band of allowed states is termed the *conduction band.* Similarly, a *valence band* of allowable energy states exists below the forbidden gap. The energy difference between the highest valence band and the lowest conduction band is the semiconductor *band-gap energy.* Charge carriers in the material cannot exist with energies in the range of the forbidden gap. The energy of this gap is one of the most important quantities that characterizes a semiconductor material. Table 1.1 presents band-gap energy values for several semiconductors of particular importance in modern MESFETs and HEMTs [2–5]. With the exception of silicon, the materials of Table 1.1 are all classified as III-V semiconductors because the constituent elements that compose the materials fall under columns III and V on the periodic table of the elements.

Table 1.1
Band-Gap Energies for Several Important Semiconductor Materials

Semiconductor Material	Room Temperature Band-Gap Energy (eV)
GaAs	1.424
AlAs	2.168
InP	1.340
$Ga_{0.47}In_{0.53}As$	0.717
$Al_{0.28}Ga_{0.72}As$	1.773
$Al_{0.47}In_{0.53}As$	1.447
Si	1.120

Some complexity in the details of the *E-k* diagram results from the solution of Schrodinger's equation. Although these details can be important for many applications (photonics in particular), the specific relationships of the *E-k* diagram can be neglected for many microwave device applications. For convenience, we develop models that consider only the band-gap energy. We commonly represent semiconductor band diagrams as illustrated in Figure 1.1.

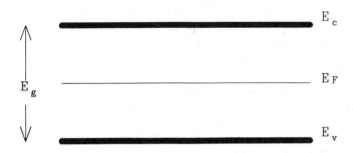

Figure 1.1 Energy band diagram of undoped semiconductor material.

In Figure 1.1, E_c and E_v represent the lowest conduction and highest valence band energies, respectively. The band-gap energy is expressed as E_g. Also shown in Figure 1.1 is the Fermi energy level, E_F. The concept of the Fermi energy level arises from statistical considerations related to available charge carriers in the material. Mathematically, it is the energy level at which the probability of finding an electron is one-half. Use of the Fermi energy level will be discussed in more detail in Section 1.1.2.

1.1.2 Doping and Carrier Concentration

A structure that comprises only semiconductor material in its purest form is not of great value for most microwave device applications. Without the ability to engineer the semiconductor properties, few of the devices utilized today would be possible.

When very small quantities of appropriate impurity atoms are added to a semiconductor crystal, the material conductivity can be altered dramatically. The addition of these impurity atoms is termed *doping*. Semiconductor crystals that resemble anything from a nearly perfect insulator to an excellent conductor are made possible by utilizing appropriate doping. Two important classes of impurity atom typically utilized to dope semiconductors are *donor atoms* and *acceptor atoms*.

The introduction of donor atoms into a semiconductor crystal results in *n*-type semiconductors. For GaAs material, Si and Se are examples of donor atoms. Likewise, both Sb and P are donor atoms for Si semiconductors. The *n*-type semiconductor is characterized by the existence of excess electrons available for current conduction in the conduction band. An allowable energy state is also created in the energy band gap very close to the conduction band edge. Small amounts of thermal energy then act to excite the surplus electrons into the conduction band where they can participate in current conduction.

Figure 1.2 illustrates an energy band diagram appropriate for *n*-type material. In the figure, the energy level created by the introduction of donor impurity atoms is expressed as E_d. This level is not typically presented in band diagrams, but appears as open circles in Figure 1.2. Notice also that the Fermi energy level is located very close to the conduction band edge. The close proximity of the conduction band edge to the Fermi level is a result of the material being doped with *n*-type impurities. The relationship between the conduction band edge–Fermi level spacing and the available free carriers can be expressed mathematically as [3, 4]

$$n = N_c \, e^{(E_F - E_c)/kT} \tag{1.1}$$

where N_c is the density of allowable energy states available in the conduction band and kT is the thermal energy of the crystal lattice. The conduction band density of states, N_c, is a quantity characteristic of the given semiconductor material. Although this value varies for different semiconductors, a value in the range of 10^{18} to 10^{20} cm^{-3} is typical of many semiconductors. The carrier concentration, n, is also commonly expressed in units of cm^{-3}, and in the fabrication of most microwave MESFETs and HEMTs, values in the 10^{14} to 10^{19} cm^{-3} range are utilized.

Figure 1.2 Energy band diagram of *n*-type semiconductor material.

Notice from equation (1.1) that if the Fermi level is located significantly below the conduction band edge (so that $E_F - E_c$ is a large negative number), the exponential portion of the expression approaches zero and the electron density is low.

Likewise, as the Fermi level gets very close to the conduction band edge, the carrier concentration becomes large.

Similar to the case for donor atoms and *n*-type material, semiconductors can be made to be *p*-type with the addition of acceptor atoms; *p*-type material is characterized by excess holes available for current conduction in the valence band. A hole is a fictitious particle assigned a positive charge equivalent to the negative of the electron charge. Although hole conduction actually results when an electron vacancy shifts position in the valence band, the concept of hole transport in semiconductors is convenient for many applications. The hole density in a semiconductor can be expressed by [3, 4]

$$p = N_v \, e^{(E_v - E_F)/kT} \tag{1.2}$$

where N_v is the valence band density of states. Figure 1.3 presents a band diagram appropriate for *p*-type material. The Fermi level for *p*-type material is shifted very near the valence band edge. This is consistent with expression (1.2) and analogous to the situation for electrons in *n*-type material.

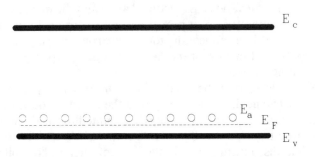

Figure 1.3 Energy band diagram of *p*-type semiconductor material.

In practice, when *n*- or *p*-type material is desired, donor or acceptor atoms are chosen that produce shallow energy levels within the band gap. The term *shallow* is used to express the close proximity of the donor level to the conduction band or of the acceptor level to the valence band edge. The use of such impurities ensures that at the temperatures of operation each impurity atom that is incorporated into the crystal produces a free carrier (hole or electron) to participate in the conduction process. For cases in which only donor dopants have been incorporated into the crystal, a formal statement of this approximation is given by

$$n \approx N_d \tag{1.3}$$

where N_d is the doping density of the donor atoms. Similarly,

$$p \approx N_a \tag{1.4}$$

for p-type material where N_a is the doping density of acceptor atoms. The number of carriers available for current conduction in the material is controlled, therefore, by the doping density of impurities incorporated into the material during the fabrication process.

Another important class of semiconductor impurity creates deep levels within the energy band gap. Such impurities cause the Fermi level to shift toward the band-gap center. According to equations (1.1) and (1.2), the electron and hole concentrations are very low. Deep levels will be discussed in more detail in Section 1.1.4.

1.1.3 Carrier Transport and Conductivity

Electrical current is determined by the rate of flow of electrical charge. Knowledge of both the amount of charge involved in the conduction process (determined by the doping density profiles and cross-sectional area) and the rate at which the charge flows (determined by the transport properties and electric field intensity) is necessary to specify current in a semiconductor. The carrier transport properties within a semiconductor are a function of the material lattice properties, lattice temperature, and the doping density.

When a free electron in the conduction band of an n-type semiconductor is exposed to an electric field, it is accelerated in the direction opposite the field vector. Conversely, a hole in p-type material is accelerated in the direction of the field. The particle continues until it suffers a collision with an ionized doping impurity or other local field discontinuity within the semiconductor. The collision is termed a *scattering event,* which has the effect of temporarily randomizing the particle's direction of movement. The electric field again begins to accelerate the particle until the next scattering event occurs. Figure 1.4 illustrates the path an electron in the conduction band of a semiconductor might travel when exposed to an electric field. This process results in a net flow of carriers in the direction determined by the electric field intensity. The current density in material dominated by electron current can be expressed as

$$J_{\text{drift}} = qnv \tag{1.5}$$

where q is the electronic charge, n is the density of free electrons in the conduction band, and v is the average velocity of the carriers in the direction determined by the electric field.

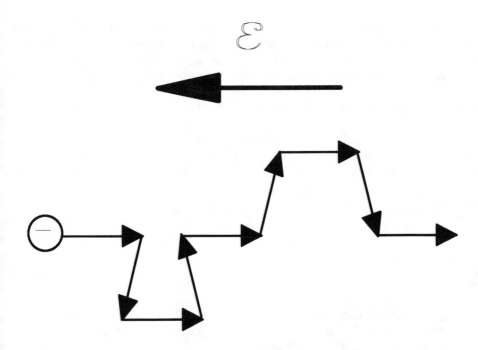

Figure 1.4 A typical path for an electron in the conduction band of a semiconductor exposed to an induced electric field.

For low electric field intensities, the average velocity of carriers is linearly proportional to the field strength. This is written as

$$v = \mu_n E \tag{1.6}$$

where E is the electric field strength, and μ_n is the low field electron mobility. Mobility is a material property and is a measure of how rapidly a charge can be moved within the semiconductor structure. Corresponding to the electron drift current and low field electron mobility in n-type material is the hole drift current and low field hole mobility for p-type material. In general, the low field mobility of a material is different for holes in the valence band than for electrons in the conduction band. The mobility is also a strong function of doping density. In GaAs and other semiconductor materials commonly utilized in the fabrication of microwave devices, electron mobility is significantly greater than hole mobility. Thus, n-type material is utilized more often than p-type material in the fabrication of these devices. For this reason, the remainder of the discussion on carrier transport will focus only on conduction band electron transport in n-type material.

Table 1.2 presents the low field mobility value of electrons in n-type material for several semiconductors of interest [2–6]. The data are presented for three different doping density levels. The mobility values for the III-V semiconductor materials are seen to be significantly higher than those for silicon for all doping density levels. In addition, note that the GaInAs material possesses mobility characteristics that are superior to those of all other materials listed.

Table 1.2
Electron Mobility for Several Important Semiconductor Materials at Various Doping Density Levels

Semiconductor Material	Electron Mobility (cm²/V-s)		
	$N_d = 10^{16}\ cm^{-3}$	$N_d = 10^{17}\ cm^{-3}$	$N_d = 10^{18}\ cm^{-3}$
GaAs	6550	4720	2735
InP	5995	3815	2125
$Ga_{0.47}In_{0.53}As$	13460	8875	4065
$Al_{0.47}In_{0.53}As$	8380	6020	3030
Si	1250	800	230

Figure 1.5 presents the low field electron mobility of GaAs as a function of doping density from 10^{13} to 10^{20} cm^{-3} [7]. The general shape of this characteristic is typical of other materials of interest. For extremely low doping density levels, low field electron mobility is not a strong function of doping density. As the doping density is increased, however, mobility values degrade rapidly.

For high electric field values, the steady-state carrier velocity becomes limited and the velocity saturates. Increasing the electric field intensity no longer results in increasing carrier velocities. Figure 1.6 shows the steady-state velocity of electrons as a function of electric field intensity for several important semiconductors doped n-type to a level of 10^{17} cm^{-3} at room temperature [3, 6]. Silicon is seen to exhibit carrier velocities far below those of the III-V materials at electric field levels below about 10 to 20 kV/cm, which is as expected from equation (1.6) and the mobility values of Table 1.2. Another feature of the velocity-field curves that distinguishes silicon is the lack of a velocity peak that exceeds the high field saturation velocity. Each of the III-V compounds shows the carrier velocity reaching a peak value at relatively low electric fields, then exhibiting a region of negative differential mobility before saturating to a final velocity value. This negative differential mobility region gives rise to the Gunn effect, which is the basis for the operation of transferred electron devices. This region may also affect the electric field and carrier concentrations observed in MESFET and HEMT devices. Finally, note that significant differences in the steady-state saturation velocity of the materials are not observed. The saturation velocity of all of the materials is close to 1×10^7 cm/s.

Figure 1.7 illustrates the effect that doping density has on the velocity-field characteristics for GaAs. As shown in Figure 1.5, the electron mobility is reduced

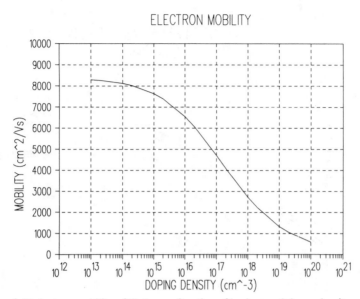

Figure 1.5 Low field electron mobility of GaAs as a function of background donor density.

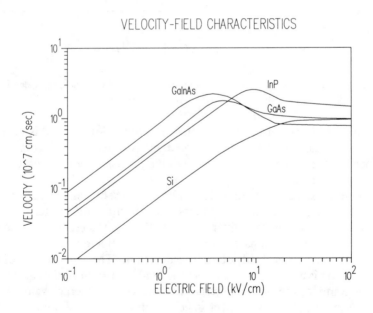

Figure 1.6 Steady-state velocity-field curves for several semiconductor materials doped n-type to a concentration of 10^{17} cm^{-3}.

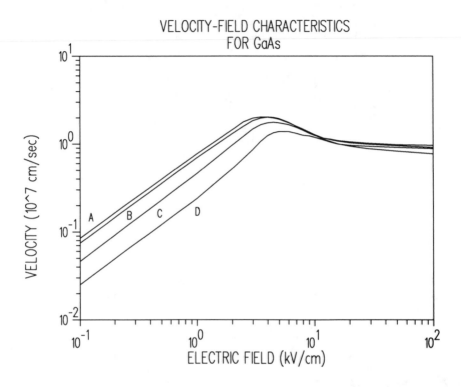

Figure 1.7 Steady-state velocity-field curves for bulk GaAs at several different doping density levels: (a) 10^{14} cm^{-3}; (b) 10^{16} cm^{-3}; (c) 10^{17} cm^{-3}; (d) 10^{18} cm^{-3}.

as doping density is increased. As expected from equation (1.6), this results in the reduced carrier velocity at low electric field levels observed in Figure 1.7. Lower peak velocity values result from increased doping density, while saturation velocity is relatively unaffected.

The data presented in Figures 1.6 and 1.7 represent steady-state properties of the semiconductors. Over short distances and for short time intervals, however, average carrier velocities can differ significantly from these steady-state values. Transient velocity phenomena can result in carrier velocities several times greater than those expected from steady-state considerations. The exact nature of the transient phenomena is a function of the electric field distribution, the free carrier profile, and the temperature of the material. Enhanced local average velocity (velocity overshoot) is greatest for abrupt electric field profiles, for low doping density levels, and for low temperatures. Although these effects are difficult to characterize directly or to simulate with extreme accuracy, transient velocity phenomena do influence the behavior of some submicron-gate-length MESFET and HEMT devices.

Figure 1.8 presents results from a Monte Carlo simulation of electron transport in GaAs material [8]. The conditions of the simulation are idealized to obtain

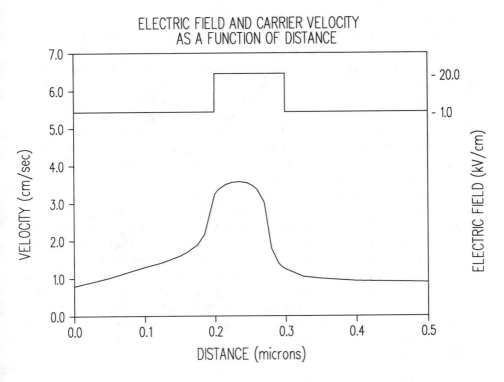

Figure 1.8 Monte Carlo predictions of transient electron velocity in GaAs for the electric field profile shown.

an upper estimate of the magnitude of velocity overshoot that might occur in a short-gate-length device. The simulated sample is 0.5 μm long, doped n-type to a carrier density level of 10^{15} cm^{-3} and is assumed to be held at room temperature. As shown in the figure, the electric field is assumed to be at a level of 1.0 kV/cm at the left edge of the sample, to step abruptly to 20 kV/cm at a distance 0.2 μm from the edge, to step abruptly back to the 1.0 kV/cm level of 0.3 μm from the left edge, and to remain at that level to the right edge of the sample. Referring to Figure 1.7, the steady-state velocity of the carriers for both 1 and 20 kV/cm is approximately 1.0×10^7 cm/s. The simulation, however, indicates that over a distance of about 0.1 μm the average carrier velocity achieves a value of over 3.0×10^7 cm/s.

The Monte Carlo predictions presented in Figure 1.8 significantly overestimate the velocity overshoot that may be realized in typical MESFET and HEMT structures. Even with this overestimation, the net average velocity across the entire structure with overshoot effects included is only about 15% higher than that predicted from steady-state considerations. In actual structures, the electric field cannot be stepped abruptly. Such a step would require infinite voltage. A more gradual and more realistic shift in electric field value, however, will reduce the velocity overshoot observed. Also, an electric field step that causes a localized acceleration of carriers will result in an accumulation of those carriers in the region just beyond the point of acceleration. These accumulated carriers will create a localized electric field that will oppose the original field. Thus, velocity overshoot becomes limited. Another important point to note is that, as doping density levels are increased, the overshoot phenomena will also become less important. A final point regarding the simulation is that the field step to 20 kV/cm was chosen to maximize the overshoot observed. Steps to either higher or lower field values result in less overall velocity enhancement. Despite these limitations to the magnitude of velocity overshoot effects, they are present in microwave devices being fabricated today, and the accurate prediction of these effects represents a challenge to device modelers.

In addition to applied electric field, several other mechanisms can act to produce current in a semiconductor structure. Applied magnetic fields, exposure to optical or other radiation sources, and thermal gradients and carrier density gradients within the material are all examples of other current-generating mechanisms. Of these other factors, for most microwave applications, only carrier density gradients are of importance. When the carrier density in a semiconductor is not homogeneous (either because of electric field or doping density distributions), carriers will tend to diffuse from regions of high carrier concentrations toward regions of lower concentrations. In a material dominated by electron current (in which hole contributions are insignificant), the current density generated by this carrier gradient can be expressed by [3, 4]

$$J_{\text{diff}} = qD_n \nabla n \qquad (1.7)$$

where D_n is the electron diffusion coefficient. For nondegenerate semiconductors, the diffusion coefficient can be related to the mobility and thermal energy by [3, 4]

$$D_n = (kT/q)\mu_n \qquad (1.8)$$

The total electron current (both drift and diffusion) is then expressed by

$$J_n = q\mu_n n + qD_n \nabla n \qquad (1.9)$$

The net effect of diffusion on carrier densities within a material is to smooth carrier density profiles. Shifts in carrier concentration levels must occur over a finite distance. In particular, semiconductor device simulations have shown that when doping density in a structure shifts abruptly from a high to a low value, the free carrier density shifts over a distance of six to seven Debye lengths [9, 10]. The Debye length of the material is given by

$$L_D = \left(\frac{\epsilon kT}{q^2 N_d} \right)^{1/2} \qquad (1.10)$$

Figure 1.9 illustrates this effect for a high-low doped structure. Although the doping of the structure shifts abruptly from a high value to a lower value in the center of the structure, the free-electron density transition is more gradual.

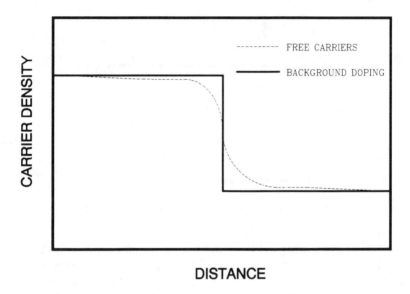

DISTANCE

Figure 1.9 Electron density at the boundary of a high-low doping junction.

1.1.4   Deep Levels

Deep levels in semiconductor material are energy states located near the middle of the band gap that can be occupied by carriers. Because carriers that occupy deep levels tend to remain trapped in those levels and unavailable for current conduction for long periods of time, deep levels are also referred to as *traps*. Both electron traps and hole traps exist. Deep levels can result from specific impurities such as boron or chrome in gallium arsenide, iron in indium phosphide, or gold in silicon. They can also result from crystal defects and damage. Other traps appear to be related to growth or fabrication procedures.

The position of traps in a given physical structure can be at interfaces (abrupt transitions in material composition or doping), at the surface of the material, or in the bulk. Figure 1.10 shows a band diagram of material dominated by traps in the bulk. Note that the Fermi level is located near mid-gap. From equations (1.1) and (1.2), this implies that few carriers are available for conduction. The incorporation of deep levels into semiconductor samples, therefore, is one method of producing semi-insulating material. If the density of deep levels is less than the density of dopant ions, the deep levels act to reduce the net free carrier concentration. For n-type material with electron traps, this can be expressed by

$$n \approx N_d - N_T \quad \text{for } N_d \gg N_T \tag{1.11}$$

where N_T is the density of electron traps.

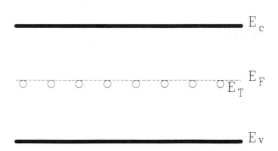

Figure 1.10 Energy band diagram for a semiconductor with excess deep levels.

In addition to affecting the free carrier concentrations, traps also affect carrier transport. Low field mobilities are reduced when deep levels are present in a semiconductor [7, 11]. Figure 1.11 presents the predicted effect of traps on mobility. The mobility is plotted for three different doping concentrations against the ratio of deep level-to-donor density. Note that the low field mobility is reduced significantly when deep level concentrations approach those of the intentional dopant atoms (i.e., as the ratio approaches one).

When deep level densities are significant in an n-type semiconductor, a high probability exists for electrons in the conduction band to fall into the traps. This is true because the traps represent a lower, more desirable energy state. While trapped, the carrier does not participate in current conduction. Once a carrier resides in a deep level, a significant amount of time (on the order of milliseconds to seconds) may be necessary for the particle to receive enough energy to move back into the conduction band. Eventually, the carrier moves into the conduction band and is directed by the applied electric field until it falls again into a lower energy trapping state.

This phenomenon has significant implications for the microwave performance of semiconductor devices. Because, at microwave frequencies, applied fields shift direction at rates significantly more rapid than the time constants associated with traps, carriers in deep levels cannot participate in the microwave properties of a device. In contrast, dc characteristics can involve carriers being alternately trapped and released. The implication here is that characteristics of devices that possess traps will shift dramatically as the stimulation is shifted from dc to high

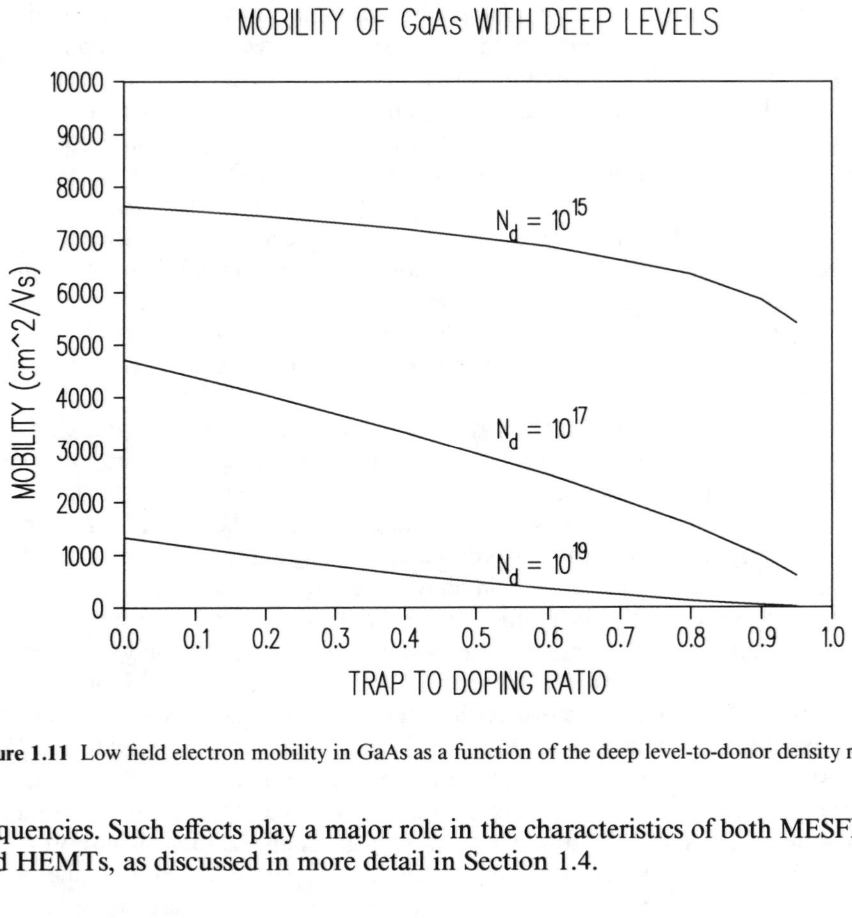

MOBILITY OF GaAs WITH DEEP LEVELS

Figure 1.11 Low field electron mobility in GaAs as a function of the deep level-to-donor density ratio.

frequencies. Such effects play a major role in the characteristics of both MESFETs and HEMTs, as discussed in more detail in Section 1.4.

1.1.5 Breakdown and Tunneling

Under certain conditions of applied voltage and at certain temperatures, the transport discussed in Section 1.1.3 no longer accurately describes the physical processes taking place in the semiconductor. Large electric field values can result in avalanche breakdown in the material. Likewise, near rectifying or heterostructure contacts (see Section 1.2), tunnel currents can become significant. Both of these phenomena can be important in determining the operational limits of MESFETs and HEMTs.

As discussed in Section 1.1.3, when the electric field intensity in a semiconductor material is increased, carrier velocity is also increased until a limiting saturation velocity is achieved. For electric field increases beyond this point, carrier velocity remains approximately constant. This is evident in all of the velocity-field curves presented in Figures 1.6 and 1.7. The additional energy associated with these electric field increases at high field levels is transferred into the crystal lattice through carrier collisions. When a carrier with some critical value of energy, E_{crit}, collides with the lattice, it frees an electron-hole pair. The colliding carrier imparts enough energy to excite an electron from the valence band into the conduction band, leaving a hole in the valence band. Both of the newly created carriers, along with the original one, are then re-accelerated by the electric field. All three carriers can then participate in other electron-hole producing collisions with the lattice. The process continues, creating an increasing number of carriers in a manner similar to an avalanche that accumulates an increasing amount of debris. Currents of very large magnitudes can be produced. Although the avalanche process itself is non-destructive, the large currents can cause permanent damage or destruction of the material lattice.

When discussing avalanche breakdown, the terms *breakdown voltage* and *critical field* are often used. The breakdown voltage is defined for a particular structure and represents the dc voltage that must be applied to a sample to observe the onset of avalanche breakdown [4, 5]. If the dimensions or the doping density profile of the structure is altered, the breakdown voltage will also shift. The critical field refers to the electric field level associated with the breakdown voltage. Although the concepts of breakdown voltage and critical field can be useful for many applications, note that avalanche breakdown is the result of carrier energy exceeding a critical value—not of applied voltage or electric field. The distinction is important at microwave frequencies because critical fields and breakdown voltages that apply for dc conditions will not, in general, be of significance at high frequencies. For a given structure, a bias voltage applied to the terminal is associated with a maximum electric field within the device. By setting the voltages appropriately, avalanche breakdown will occur. If these same conditions are achieved for only a short period of time, however (by pulsing the bias on briefly or as part of a high frequency signal swing), the carriers in the material may not achieve sufficient energy to produce the avalanche effect [12]. Thus, the dc breakdown characteristics of a semiconductor device such as a MESFET or HEMT are not typically similar to the RF breakdown characteristics.

A second type of breakdown that can impose performance limits on MESFETs and HEMTs is *Zener breakdown* [4, 5]. Zener breakdown occurs when a significant number of carriers *tunnel* through an energy barrier in a *p-n* junction or Schottky contact (see Section 1.2). Tunneling is a phenomenon that results from quantum mechanical considerations. For extremely thin barriers (less than approximately 100 Å), carriers may tunnel through an energy barrier without having sufficient energy to surmount the barrier. Three ingredients are essential to produce significant tunneling:

1. The barrier must be thin. The probability of tunneling decreases rapidly with increased thickness.

2. A high density of free electrons must exist on one side of the barrier.
3. A high density of unoccupied energy states available for electron occupancy must exist on the other side of the barrier.

These conditions are achieved for highly doped ($\approx 10^{18}$ cm^{-3} or greater) *p-n* or Schottky barrier junctions. Both MESFET and HEMT scaling rules result in Schottky barrier gate terminals fabricated on highly doped material for short-gate-length devices.

1.2 CONTACT PROPERTIES

Connections between the bulk semiconductor and other electrical components or equipment are established via a number of different types of semiconductor contacts. The properties of these contacts are as critical to overall device performance as the properties of the semiconductor. In addition to the contacts made between bulk semiconductor and electrical terminals, contacts between semiconductors of different doping types or different molecular constituents are also important.

1.2.1 Ohmic Contacts

The electrical characteristics of the ideal ohmic contact are purely resistive in nature. This means that the current through the contact is linearly proportional to the voltage drop across it, so that it can be modeled as a linear resistor. This ohmic behavior results from the fact that the junction offers negligible resistance and, hence, the junction I-V characteristics are dictated by the bulk semiconductor properties as described in Section 1.1. Ohmic contacts are critical to the operation of all semiconductor devices, as they provide a "node" through which signals can traverse with minimal resistance and no rectification. The drain and source contacts of a MESFET or HEMT are always ohmic and have associated with them parasitic resistances R_d and R_s that model the contact ohmic loss and the losses in the bulk semiconductor.

Theoretically, ohmic contacts are formed by joining an *n*-type semiconductor with some work function ϕ_S to a metal with a smaller work function ϕ_M, i.e., $\phi_S > \phi_M$. This situation is depicted in Figure 1.12. The Fermi levels are aligned by the transfer of electrons from the metal to the semiconductor. This electron transfer raises the semiconductor electron energies relative to the metal at equilibrium. The barrier to electron flow is thereby diminished, permitting electrons to flow across the barrier with minimal resistance. Because the potential difference required to align the Fermi levels is provided by majority carriers, in this case, no depletion region forms.

In practice, finding a metal for which the band diagram of Figure 1.12 applies is difficult. The existence of a high density of surface states located within the band

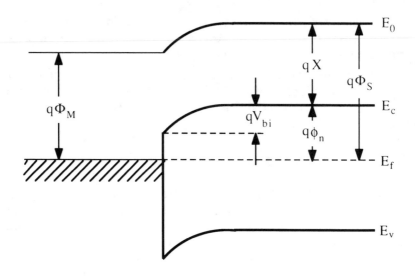

Figure 1.12 Band diagram for an idealized ohmic contact.

gap on the semiconductor surface causes the Fermi level to be "pinned" regardless of the metal work function. Real ohmic contacts are made by heavily doping the semiconductor region adjacent to the contact. If the doping is sufficiently high, such that the semiconductor depletion thickness is on the order of tens of angstroms, electrons can tunnel through the barrier rather than having to surmount it. This process results in a very small amount of resistance and, therefore, a good ohmic contact. The tunneling process is illustrated in Figure 1.13. For a triangular barrier with height E_g and width L, the tunneling probability can be approximated by [5]

$$\theta = e^{(-qBL/E_g)} \tag{1.12}$$

where q is the electron charge and

$$B = \frac{4\sqrt{2m^*}E_g3/2}{(3q\hbar)} \tag{1.13}$$

where m^* is the effective mass of an electron and $\hbar$ is the reduced Planck constant. Note that according to equation (1.12) tunneling probability increases exponentially with either a decrease in barrier width L or an increase in electric field (doping).

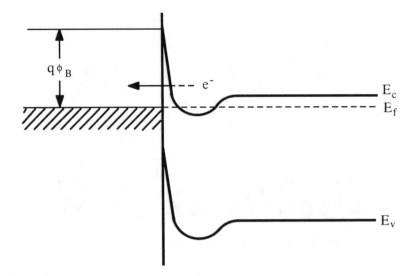

Figure 1.13 Ohmic contact band diagram illustrating the tunneling process.

1.2.2 The *p-n* Junctions

While an ohmic contact is useful for providing a minimal resistance path to and from a bulk semiconductor, the various functions that solid-state devices perform (e.g., rectification, amplification, switching, *et cetera*) require the use of non-ohmic contacts such as *p-n* junctions or Schottky barriers. Although Schottky barriers rather than *p-n* junctions are used in MESFETs and HEMTs, the development of the Schottky barrier equations follows as a special case of the material presented here. The presentation is largely qualitative, although the relevant expressions needed for modeling and parameter extraction are given. The objective here is to relate the device physics to relevant modeling expressions.

Shown in Figure 1.14 are the three basic modes of operation of a *p-n* junction. The figures present a simplified account of the physical processes that take place in a *p-n* junction. We assume that the bias is applied to the bulk *p* and *n* regions through ohmic contacts with negligible resistance. Hence, the entire applied potential appears across the junction transition region. In addition, an abrupt change in doping between the *p* and *n* region semiconductors is assumed. Before the *p*- and *n*-type semiconductors are joined, the *n* material has a high concentration of electrons and the *p* material has a high concentration of holes. When the two regions are joined, the carriers begin to diffuse into the adjacent semiconductor. As this happens, the two regions become depleted of their respective carriers near the interface and leave behind uncompensated donors (*n* region) and acceptors (*p* region).

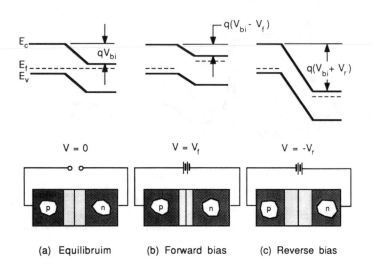

Figure 1.14 The p-n junction transition region width and band diagram for three biasing conditions: (a) equilibrium; (b) forward bias; (c) reverse bias.

This sets up an electric field that opposes further buildup of the diffusion current I_{diff}. The drift current, I_{gen}, is comprised of electron-hole pairs generated near the space charge region boundary that are able to diffuse into the space charge region and are subsequently swept down the potential barrier by the electric field. At equilibrium, the total device current is given by:

$$I = I_{\text{diff}} - |I_{\text{gen}}| = 0 \tag{1.14}$$

where the currents I_{diff} and I_{gen} are equal in magnitude but oppositely directed in the device.

Under an applied bias, the junction is no longer in equilibrium, resulting in a separation of the Fermi levels and raising or lowering of the potential barrier. Under forward bias, the electrostatic potential of the p side is raised with respect to the n side, thereby lowering the potential barrier. Because diffusion current is comprised of electrons and holes that have sufficient energy to surmount the potential barrier and diffuse to the opposite side of the junction, diffusion current rises dramatically under forward bias. In fact, the diffusion current under forward bias V_f is increased by a factor $e^{qV_f/kT}$ over its equilibrium value I_{diff}.

Under reverse bias conditions, the barrier is greatly increased such that virtually no majority carriers have sufficient energy to diffuse to the opposite side of the junction. The diffusion current for reverse bias V_r is reduced by the factor

$e^{-qV_r/kT}$. If the bias V is applied, then the nonequilibrium case can be extended by including the exponential factor in I_{diff}:

$$I = |I_{\text{gen}}|(e^{qV/kT} - 1) \qquad (1.15)$$

which is the familiar diode equation.

Equation (1.15) assumes that no recombination occurs in the space charge region. In reality, some recombination does take place, and the effect on (1.15) is a decrease in the exponent. The diode current is corrected for this space charge recombination using the *ideality factor n*:

$$I = I_S(e^{qV/nkT} - 1) \qquad (1.16)$$

where I_S is the reverse saturation current. The expression for current is now given in terms of two semiempirical parameters, which can be used for modeling diode I-V data. The specific procedure for extracting the parameters for a best fit to the measured data will be the subject of Sections 3.1 and 4.1.

Under high current conditions, most diodes begin to deviate from the ideal linear behavior (as seen on a log scale) predicted by equation (1.16) and as shown in Figure 1.15. Ohmic losses in the external contacts and in the neutral regions are responsible for this deviation. These losses can usually be accounted for adequately with a single series resistance. For the case of a Schottky barrier, the ohmic loss is a metalization resistance and is handled in the same fashion.

In addition to the current-voltage characteristics, capacitance-voltage behavior of the diode is also of interest. Two types of capacitances exist in *p-n* junctions: (1) junction capacitance, which results from uncompensated donor and acceptor ions in the space charge region and dominates under reverse bias conditions, and (2) charge storage capacitance, which, as the name implies, results from charge storage effects under forward bias conditions. When carriers have sufficient energy to overcome the potential barrier and diffuse to the other side of the junction, they become minority carriers (i.e., electrons in the *p*-type material). As such, they eventually recombine with the majority carriers with which they are surrounded. This recombination process has associated with it a characteristic time, τ_p for holes and τ_n for electrons. The stored charge is then $Q_p = I_{dn}\tau_p$ for holes and $Q_n = I_{dp}\tau_n$ for electrons, where I_{dn} and I_{dp} are the diffusion currents associated with electron injection and hole injection, respectively. The charge storage capacitance, or "diffusion capacitance" as it is sometimes referred to, is then given by the derivative of charge with respect to applied voltage.

The reverse bias capacitance characteristics of a *p-n* or Schottky junction are of primary concern because the normal mode of operation in a MESFET has the gate reverse biased with respect to the source. Referring to the idealized *p-n* struc-

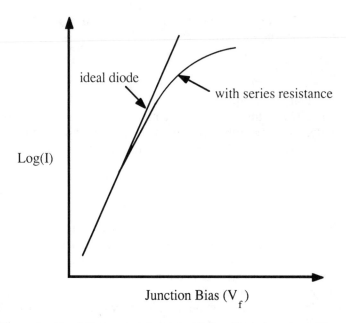

Figure 1.15 The *p-n* junction I-V characteristics under ideal and high current conditions.

ture of Figure 1.16, the reverse bias capacitance characteristics of the *p-n* junction can be developed as follows: Assuming an abrupt depletion region, the width of the transition region can be expressed as

$$W = \left[\frac{2\epsilon(V_{bi} - V)}{q} \left(\frac{1}{N_a} + \frac{1}{N_d} \right) \right]^{1/2} \tag{1.17}$$

from which the junction capacitance is found as $C_j = \epsilon A/W$, where A is the junction cross-sectional area. For a p^+-n junction, the expression for the junction capacitance simplifies to [4]:

$$C_j = \frac{A}{2} \left(\frac{2q\epsilon}{V_{bi} - V} N_d \right)^{1/2} \tag{1.18}$$

which is also the correct expression for a Schottky junction on *n*-type material.

Equation (1.18) expresses the junction capacitance as a function of the physical properties of the device and the applied bias. Although (1.18) is strictly valid only for uniform doping, an effective value of N_d or a modified exponential term

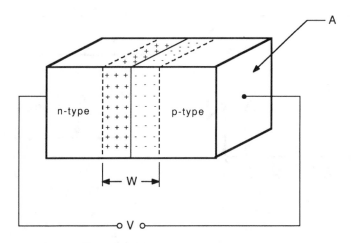

Figure 1.16 Idealized *p-n* junction structure.

can be used to model devices with nonuniform profiles. For circuit modeling, the following form of (1.18) is often used:

$$C_j = \frac{C_{j0}}{\left(1 - \dfrac{V}{V_{bi}}\right)^m} \tag{1.19}$$

where C_{j0} is the zero-bias junction capacitance and m depends on the doping profile of the diode. Typically, equation (1.19) would be used by extracting values of C_{j0} and m for a best fit to measured capacitance data over bias.

1.2.3 Schottky Barrier Junctions

Metal-semiconductor rectifying contacts or Schottky barriers consist of a thin layer of metal deposited on a semiconductor. The band characteristics for a metal-semiconductor (*n*-type) junction in equilibrium are shown in Figure 1.17. Because ϕ_S > ϕ_M (the usual case for a metal-semiconductor contact), the average total energy of electrons in the semiconductor is greater than that in the metal, resulting in a transfer of electrons to the surface of the metal. The charge on the metal side resides on an infinitesimally thin layer, while the width of the positively charged transition region in the semiconductor is a function of the doping density. The behavior of the junction is analogous to that of a *p-n* junction with an infinitely doped *p*-region (one-sided step junction).

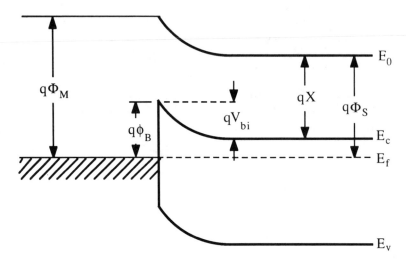

Figure 1.17 Ideal metal-semiconductor band diagram.

The barrier from semiconductor to metal varies with bias, giving rise to the Schottky barrier I-V characteristics. When the metal is biased positively with respect to the semiconductor, the barrier is lowered, and the probability of an electron overcoming the barrier and moving to the metal is increased by the same exponential factor as in the *p-n* junction. Under reverse bias, the barrier increases, and the probability of an electron moving into the metal decreases by the same exponential factor.

Equations (1.16) and (1.19), developed for the *p-n* junction, are applicable to Schottky barriers as well and are often used in modeling the I-V and C-V characteristics of the Schottky gate junction in FETs. Finally, because minority carriers are not involved in the current transport, no diffusion capacitance is associated with the forward bias operation and therefore its switching speed is superior to that of the *p-n* junction.

1.2.4 Heterostructures

A *heterostructure* or *heterojunction* occurs when contact is made between two semiconductor materials with different band-gap energies. The differences between the band-gap energies can be used to advantage in the fabrication of certain types of devices. These devices include semiconductor lasers, LEDs, heterostructure bipolar transistors (HBTs), and HEMTs. Heterostructures formed between GaAs and AlGaAs and between InGaAs and AlGaAs are of particular interest for the HEMT applications discussed in this book. HEMT devices based on InP and related III-V semiconductors are also of significant importance for many applications.

Heterojunction devices present difficult challenges to both device fabrication and device modeling engineers. The fabrication process requires the doping density, layer thickness, and material composition to be altered abruptly during growth of the device structure. Extremely thin heterostructure layers (on the order of 50 Å thick or less in some cases) are required in modern HEMT devices. Doping, layer thickness, and composition must all be very tightly controlled to produce a useful device. In addition to these concerns, fabrication of useful heterostructures also requires consideration of material lattice constants, thermal expansion coefficients, and interface states.

The heterostructure also adds complexity to the physical description required to predict device behavior. Band-gap discontinuities affect current flow and electric field distributions in first-order ways. Quantum mechanical effects are also often of first-order importance. All of these phenomena must be accounted for in physically based device models.

Figure 1.18 presents the band diagram for a *p-n* heterojunction. The *p*-type material depicted in the figure has a smaller band gap than the *n*-type material. The

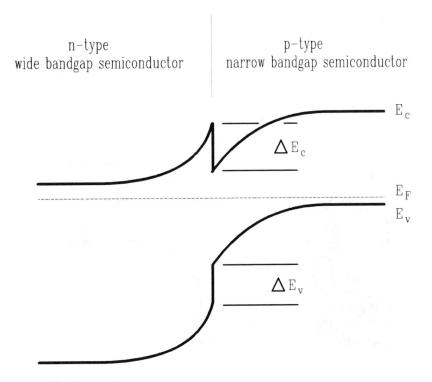

Figure 1.18 Band diagram for a *p-n* heterojunction.

band diagram can be compared to the *p-n* junction diagram for the homogeneous material composition shown in Figure 1.14(a). For the heterojunction case, band-gap discontinuities exist in both the conduction and valence band. In Figure 1.18, these discontinuities are labeled "ΔE_c" and "ΔE_v," respectively. The amount of conduction and valence band discontinuity for two materials forming a hetero-junction is determined by a number of factors, including the band-gap energies and electron affinities of each of the materials.

Physically based analytical expressions can be developed to describe the current flow and capacitance across a *p-n* heterojunction as a function of applied voltage [4, 13]. Although such an expression is useful in the analysis of HBTs, it is not easily applied to HEMT devices and, therefore, is not pursued here.

Figure 1.19 presents the band diagram for a Schottky barrier placed on a semiconductor heterostructure typical of the structures exploited for HEMT fab-

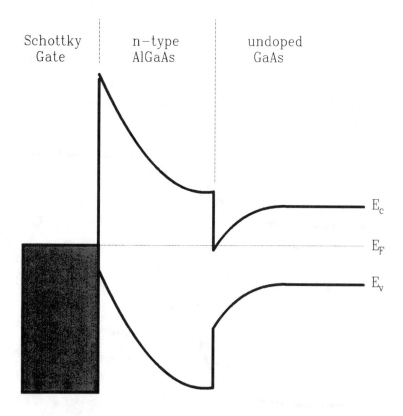

Figure 1.19 Band diagram for a Schottky contact made to a heterostructure typical of fabricated HEMT devices.

rication. The wide-band-gap material is doped *n*-type, but is depleted of free carriers by the reverse or zero-biased Schottky contact. The narrow-band-gap material shown on the right portion of the figure is lightly doped *p*-type. The most important feature of the band diagram related to HEMT operation occurs at the heterojunction interface between the two materials. At this boundary, the band-gap discontinuities cause the conduction band of the narrow-band-gap material to dip below the Fermi level. From equation (1.1), this implies that the free carrier concentration is very high where this dip occurs. Note that the region of high carrier concentration is extremely thin. This thin region of high carrier density is the primary property of the heterostructure exploited in the fabrication of HEMTs and will be discussed in greater detail in Section 1.5.

1.3 MESFET DESCRIPTION

The cross section of a MESFET device is illustrated in Figure 1.20. Three metal electrode contacts are made to a thin semiconductor active channel layer. The contacts are labeled "source," "gate," and "drain." For microwave and millimeter wave applications, the thin active layer is almost always *n*-type GaAs material. The active channel is fabricated by utilizing ion implantation of donor atoms into semi-insulating material or epitaxial growth of doped material. Doping concentrations are often varied intentionally with depth to achieve specific performance. Regions of highly doped material beneath the source and drain contacts such as those shown in the figure are often incorporated into the structure to reduce unwanted contact resistance. These regions are produced using an additional implant or diffusion process. The active channel lies on top of semi-insulating GaAs.

Figure 1.21 presents another perspective of the MESFET device with several important geometric dimensions labeled. The most important dimension that characterizes the MESFET physical structure is the gate length, which is labeled "L" in Figure 1.21. This dimension is critical in determining the maximum frequency limits for MESFET devices. The gate width, Z, is another physical device dimension that is of primary importance in the determination of device behavior. In fact, MESFETs are often described only in terms of the gate dimensions. A device is referred to as a 0.5×300-μm device when the gate length is 0.5 μm and the gate width is 300 μm. Other characteristic dimensions include the epi-thickness, a; the gate-to-source and gate-to-drain terminal spacings, L_{gs} and L_{gd}; and the drain and source terminal lengths, L_d and L_s. A microwave or millimeter wave device typically has a gate length in the 0.1- to 1.0-μm range. The epi-thickness is approximately 0.2 to 0.3 times the gate length dimension. Terminal spacings are on the order of one to four times the gate length. The gate width can vary significantly, ranging from about 100 to 2000 times that of the gate length. Note that the device current is directly proportional to gate width because the cross-sectional area available for channel current is proportional to Z. For low-noise, low-current applica-

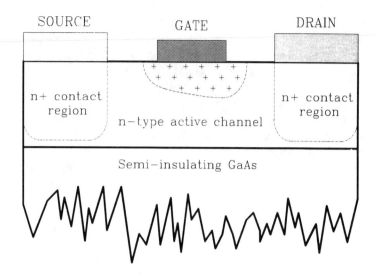

Figure 1.20 Cross-sectional view of a MESFET.

tions, therefore, relatively small-gate-width devices are utilized. In contrast, large-gate-width devices are typically used in power applications.

Actual layout of MESFET terminals is more complex than implied by Figures 1.20 and 1.21. Connections must be made to all three of the terminals—source, gate, and drain. These connections are either bond wires or additional metal layers, and they require that large metalized areas be included as part of the structure. Figure 1.22 illustrates two common approaches to the MESFET ter-

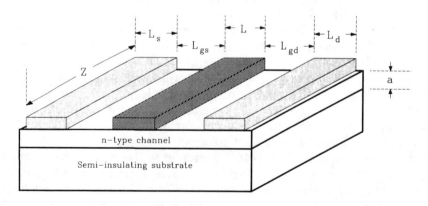

Figure 1.21 Important dimensions in the MESFET.

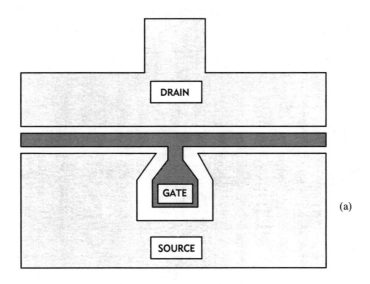

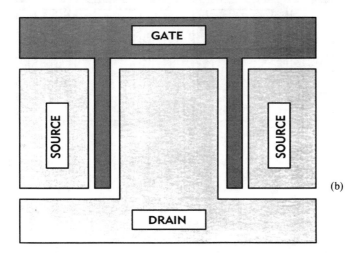

Figure 1.22 Examples of terminal layouts used to fabricate MESFETs: (a) a "T-shaped" gate MESFET; (b) a standard cell MESFET.

minal layout problem. These two approaches are easily extended to realize larger gate width devices by repeating the patterns presented, as illustrated in Figure 1.23.

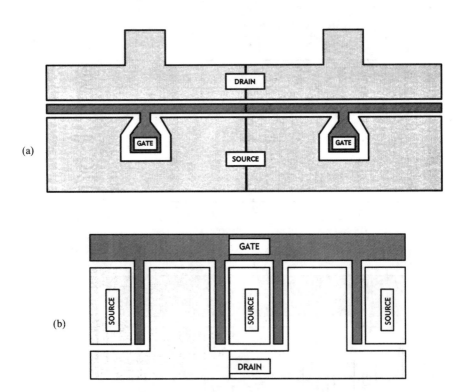

Figure 1.23 Extensions of the layout patterns illustrated in Figure 1.22 to achieve larger gate widths: (a) a "pi-gate" MESFET; (b) repeated standard cells.

1.3.1 Current-Voltage Characteristics

Both the source and drain terminals are ohmic contacts as described in Section 1.2.1. Most microwave MESFETs are *depletion mode* devices. This means that in the absence of applied reverse gate bias, current can flow between the source and drain contacts. *Enhancement mode* devices do not conduct current between the drain and source unless forward gate bias is applied. For depletion mode devices, when reasonably low bias potentials are applied between the source and drain contacts, a current flows through the channel. This current is linearly related to the voltage across the terminals. Thus, for small biases, the source-drain terminals behave similar to a linear resistor. For higher drain-source bias levels, the electrons in the semiconductor material will attain their maximum carrier velocity as described in Section 1.1.3 and illustrated in Figures 1.6 and 1.7. Associated with the saturation of carrier velocity is the saturation of channel current. This relationship between velocity and current is expressed in equation (1.5).

The gate contact in a MESFET device is a Schottky barrier (see Section 1.2.3). The energy band bending produced by making Schottky barrier contact with the semiconductor creates a layer beneath the gate that is completely depleted of free charge carriers. As no free carriers exist in this depletion layer, no current can flow through it. The available cross-sectional area for current flow between the source and drain is reduced by the existence of this depletion layer. As reverse bias is applied to the gate, the depletion layer penetrates deeper into the active channel. These further reductions in cross-sectional area result in further current reductions. The gate bias, then, acts as a mechanism for limiting the maximum amount of source-drain current that can flow. When enough reverse bias is applied, the depletion region will extend across the entire active channel and allow essentially no current to flow. The potential required to accomplish this is termed the "pinch-off potential" or "pinch-off voltage."

The current-voltage relationships expected from an ideal MESFET as described above are illustrated in Figure 1.24(a). The channel current is plotted as a function of applied drain-source potential for several different gate-source voltage levels. The saturated current level achieved when $V_{gs} = 0$ is commonly expressed as I_{dss}. Saturated current levels for other gate-source bias conditions are then expressed in terms of %I_{dss}. The magnitude of the current for a given drain-source bias will be directly proportional to the gate width of the device. For that reason, drain current at a particular bias point is often discussed in terms of the current per unit gate width, I_{ds}/Z.

The current-voltage characteristics of actual MESFET devices are similar to these ideal characteristics, with the important exception that the slope of the curves remains slightly positive even after semiconductor-limited velocity has been reached. Figure 1.24(b) presents measured I_{ds}–V_{ds} characteristics for two GaAs MESFETs. The two devices have been fabricated on identical material. All device dimensions are equivalent with the exception of the gate length. The gate width of both devices is 300 μm.

Note that if optimal device performance is desired, decreases in gate length should be associated with corresponding decreases in other device dimensions as well as increases in channel doping densities. This process is termed *scaling* and will be discussed in greater detail in Chapter 5. The 1×300- and 0.5×300-μm device characteristics presented in this section are utilized only to indicate the effect a single parameter, gate length, has on device characteristics. The 0.5-μm gate length device fabrication process has not been optimized.

Although the curves of Figures 1.24(a) and 1.24(b) are similar in many respects, the differences between the characteristics have significant implications with regard to device performance. Not only do the actual measured characteristics exhibit a finite slope, but the slope is greater for the 0.5-μm gate length device than

IDEALIZED MESFET I-V CHARACTERISTICS

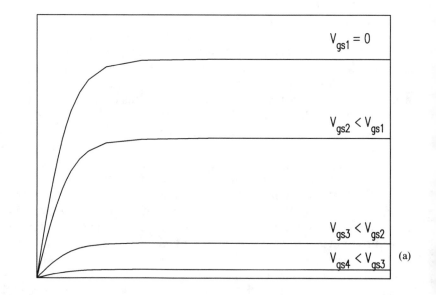

DRAIN-SOURCE VOLTAGE

Figure 1.24 The dc current-voltage characteristics of MESFETS: (a) ideal drain-source current as a function of drain-source voltage for several gate-source voltage levels; (b) measured drain-source current as a function of drain-source voltage for several gate-source voltage levels. Curves for both a 1-μm gate length device and a 0.5-μm gate length device are compared; (c) measured drain-source current as a function of gate-source voltage for a drain-source bias of 5.0 V. Curves are presented for both a 1-μm gate length device and a 0.5-μm gate length device.

for the 1.0-μm gate length device. Figure 1.24(c) presents the drain current dependence on gate-source voltage and illustrates that the device pinch-off voltage is increased in magnitude for the shorter gate length device. These short channel effects are often more pronounced in devices with extremely small gate lengths (typically with gate lengths below 0.5 μm) or improperly scaled devices. The effects sometimes lead to severely degraded device performance. Small gate length-to-channel thickness ratios, L/a, and carrier injection into the semi-insulating substrate are two of the causes of short channel effects.

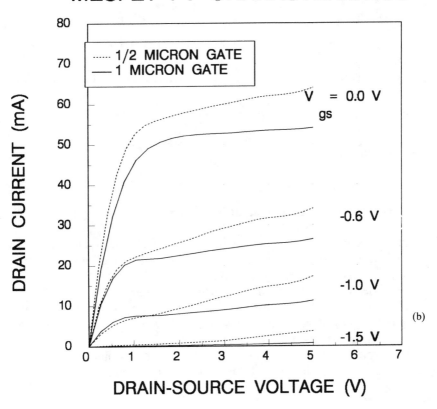

Figure 1.24 continued

MESFET I-V CHARACTERISTICS

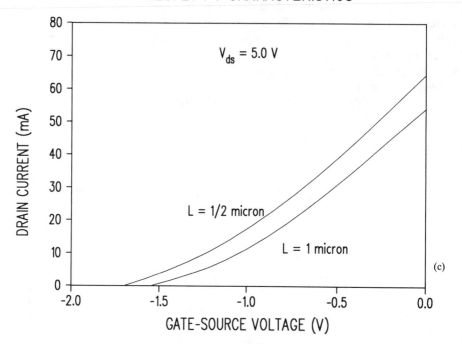

Figure 1.24 continued

1.3.2 Transconductance and Output Resistance

For analog applications, the plots of current-voltage characteristics such as those of Figure 1.24 do not reveal as much about device performance as do the derivatives of these curves. Both device output conductance and transconductance are defined in terms of derivatives of the curves presented in Figure 1.24.

The derivative of the drain-source current with respect to drain-source voltage while gate-source voltage is held constant is defined as the output conductance of the device. Note that the output characteristics of the device are often more conveniently expressed in terms of output resistance. These quantities are related simply as the inverse of each other. Mathematically, the output conductance and resistance are given by

$$g_{ds} = \frac{1}{r_{ds}} = \left. \frac{dI_{ds}}{dV_{ds}} \right|_{V_{gs} = \text{ constant}} \qquad (1.20)$$

The output conductance of the device is an important characteristic in analog applications. It plays a significant role in determining the maximum voltage gain attainable from the device and is extremely important for determining optimum output matching properties. In general, for a device to have a low value of output conductance is desirable, or, equivalently, an extremely high output resistance.

Figure 1.25 presents measured microwave output resistance for a 0.5×300-μm device. As expected from Figure 1.24(b), the output resistance is low at low drain-source bias levels and increases dramatically as the device reaches saturation. This is true for all of the curves presented except for a gate bias level of -1.5 V. At this bias level, the active channel is nearly pinched off by the depletion region. Second-order conduction mechanisms including injection of charge into the non-ideal semi-insulating substrate and conduction via deep levels at the surface and channel-substrate interface begin to dominate. These same effects contribute to causing finite output resistance even in the current saturation region.

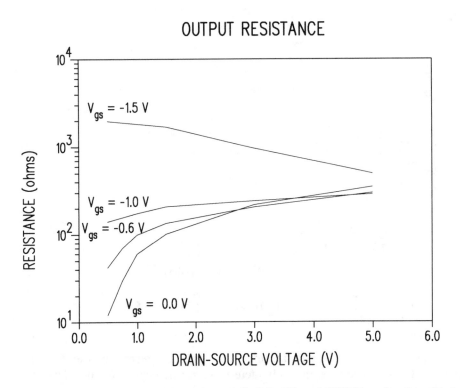

Figure 1.25 Measured microwave output resistance of a 0.5×300-μm MESFET as a function of drain-source voltage.

Device dimensions and channel material properties both affect output resistance of the MESFET. As in the case of channel current, the magnitude of the device output conductance (the inverse of the resistance) is directly proportional to device gate width. Figure 1.26 presents measured microwave output resistance for both the 1- and 0.5-μm gate length devices used to produce Figures 1.24(b) and 1.24(c). The data illustrate the relationship between gate length and output resistance, which is expected from examination of the I-V characteristics of Figure 1.24(c). For devices with equivalent gate widths, short gate lengths typically result in lower output resistances. The output resistance can also be reduced by increasing channel doping concentrations, N_d, or the device epi-thickness, a.

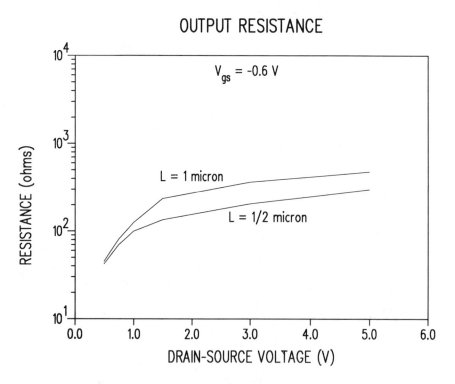

Figure 1.26 A comparison of measured microwave output resistance of MESFETs with 1- and 0.5-μm gate lengths.

Because deep levels play an important role in determining the output resistance of GaAs MESFETs, these characteristics are strongly frequency dependent at relatively low frequencies (i.e., dc to approximately 1 MHz). Thus, a plot of dc measured output resistance of the device is functionally similar to the microwave characteristics of Figure 1.25, but is shifted to higher values of resistance when the device is in saturation. This problem will be discussed in more detail in Section 1.4.

The device transconductance is defined as the slope of the I_{ds}–V_{gs} characteristics with the drain-source voltage held constant. The mathematical statement of this is

$$g_m = \left.\frac{dI_{ds}}{dV_{gs}}\right|_{V_{ds}=\text{constant}} \tag{1.21}$$

The transconductance of the device is one of the most important indicators of device quality for microwave and millimeter wave applications. When all other characteristics are equal, a device with high transconductance will provide greater gains and superior high-frequency performance.

Figure 1.27 presents the measured microwave transconductance of a 0.5×300-μm gate length device as a function of gate-source voltage. The charac-

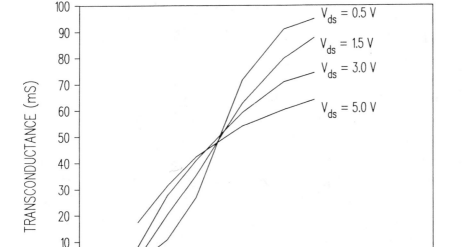

Figure 1.27 Measured microwave transconductance of a 0.5×300-μm MESFET as a function of gate-source voltage.

teristics presented in the figure are in agreement with what is expected from Figure 1.24(c) and equation (1.21). The transconductance is essentially zero for gate bias levels below the pinch-off voltage. As the reverse bias on the gate is reduced toward zero, the transconductance increases monotonically. This is characteristic of GaAs MESFET behavior and one of the key performance traits that distinguishes MESFETs from HEMTs. This difference is discussed in more detail in Section 1.5.

The device transconductance is greatly affected by device dimensions and channel material properties. Figure 1.28 presents measured device transconductance for the two devices with characteristics as shown in Figures 1.24(b) and 1.24(c). The 0.5-μm device is seen to have a significantly improved transconductance over the 1-μm device. When other device dimensions and the channel doping density are properly scaled with gate length (see Section 5.3), this observed trend is found to be true. As with channel current and output conductance, MESFET transconductance is directly proportional to gate width. For this reason, comparisons between devices are often made by examining the transconductance per unit gate width, g_m/Z.

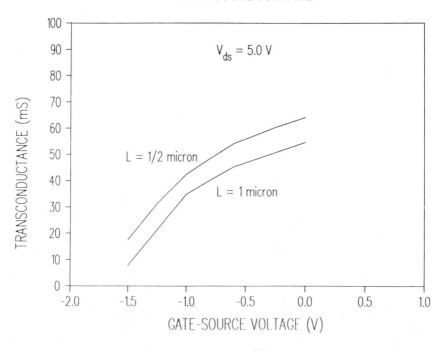

Figure 1.28 A comparison of measured microwave transconductance of MESFETs with 1- and 0.5-μm gate lengths.

MESFET transconductance is influenced—although to a much lesser degree than the output conductance—by the presence of deep levels in the device structure. Measured microwave transconductance is typically 5 to 25% lower than measured dc transconductance for a GaAs MESFET. A more detailed discussion of this topic is presented in Section 1.4.

1.3.3   Capacitance-Voltage Characteristics

As reviewed in the previous discussions, the extent and dimensions of the depletion region beneath the gate of the MESFET are determined by the bias applied to the device terminals. Both the gate-to-source potential and gate-to-drain potential affect the charge distribution in the channel.

Figure 1.29 illustrates three different depletion region shapes that can be realized in the device. Figure 1.29(a) represents the symmetric bias case in which the gate-to-drain bias is equal to the gate-to-source bias. The illustration also assumes that the gate terminal is located directly in the middle of the gap between the source and drain terminals. The resulting depletion region is symmetric with respect to the source and drain. The case in which the gate-drain reverse bias is greater than the gate-source reverse bias is illustrated in Figure 1.29(b). For these circumstances, the depletion region is deeper at the drain end of the gate than at the source end of the gate, and it extends closer to the drain than to the source. This situation represents the normal MESFET operating region in most amplifier, mixer, and oscillator applications. Figure 1.29(c) represents the case in which the gate-source reverse bias is greater than the gate-drain reverse bias. This situation is the mirror image of the case shown in Figure 1.29(b).

Because the charge is redistributed with changing gate-source and gate-drain voltages, two important capacitances can be defined. The general definition of capacitance is given by

$$C = \frac{dQ}{dV} \qquad (1.22)$$

where Q is the total charge involved in transition and V is the applied voltage. This definition is easily applied to two terminal devices such as parallel plate capacitors or simple p-n junctions. Because in a MESFET only one region of depletion charge is shared between the gate-source and gate-drain capacitances, these quantities must be defined more carefully. The gate-source capacitance may be defined as

$$C_{gs} = \frac{dQ_g}{dV_{gs}} \bigg|_{V_{gd}=\text{constant}} \qquad (1.23a)$$

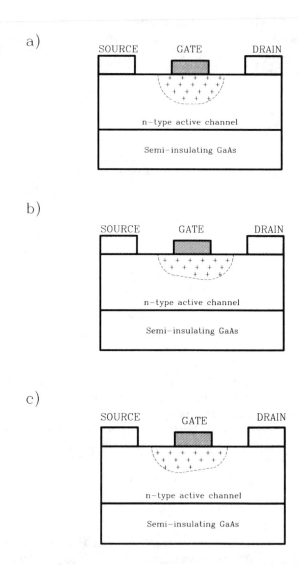

Figure 1.29 Depletion region shapes for various applied bias levels: (a) gate-source voltage is equal to gate-drain voltage; (b) reverse bias applied to gate-drain is greater than the bias applied to gate-source; (c) reverse bias applied to gate-source is greater than the bias applied to gate-drain.

In equation (1.23a), Q_g is the depletion region charge beneath the gate. Similarly, the gate-drain capacitance can be defined as

$$C_{gd} = \frac{dQ_g}{dV_{gd}} \bigg|_{V_{gs}=\text{constant}}$$ (1.23b)

The definitions expressed by equations (1.23) are not the only possible definitions applied to the MESFET capacitance problem. Under typical operating conditions, the MESFET source terminal is grounded. A negative bias is applied to the gate and a positive bias to the drain. Therefore, it is the gate-source and drain-source dc bias voltages that are directly controlled. For this reason, the gate-source capacitance is often defined as

$$C_{gs} = \frac{dQ_g}{dV_{gs}} \bigg|_{V_{ds}=\text{constant}}$$ (1.24)

Equations (1.23a) and (1.24) are not equivalent, but describe slightly different quantities. The distinction is usually minor, but can be significant if calculations are based on a physically based model. When capacitance is determined by measurements or derived from empirical models, however, the capacitance definitions given by equations (1.23) and (1.24) are not applied. Instead, the capacitance values are defined in terms of an equivalent circuit and the values required of the equivalent circuit elements to predict device behavior accurately. Appropriate equivalent circuit topologies are discussed in Chapter 2.

The gate-source capacitance is important for microwave applications because it has a significant impact on both device input impedance and ultimate frequency performance. To first order, the input impedance of a MESFET in a standard common source configuration, with typical bias levels applied, is simply the impedance of the gate-source capacitance in series with a few ohms of resistance. At high enough frequencies, the gate-source capacitance represents nearly a short circuit. The higher the C_{gs} value, the lower the frequency at which this situation occurs. At such frequencies, the device will not produce useful gain. To increase the high-frequency characteristics of GaAs MESFETs, therefore, the reduction of gate-source capacitance is desirable.

Figure 1.30 presents measured gate-source capacitance for a 0.5×300-μm device. The figure shows capacitance increasing monotonically as gate-source voltage is increased from pinch-off toward zero bias. This general trend is true for all drain-source bias levels and is typical of MESFET performance. The characteristics observed for this device are nearly linear with gate-source voltage. This linear relationship is not unusual, but is not common to all MESFETs. Some devices exhibit gate-source capacitance characteristics that increase at greater than linear rates.

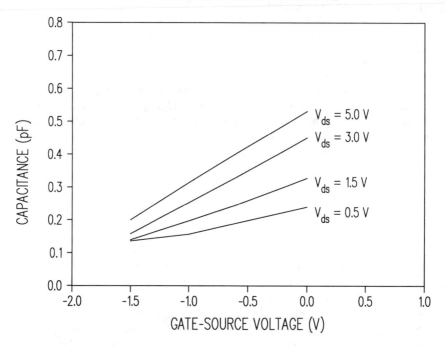

Figure 1.30 Measured microwave gate-source capacitance of a 0.5×300-μm MESFET as a function of gate-source voltage.

As the gate dimensions of a device are altered, the depletion region underneath the gate is also changed. This directly affects gate-source capacitance. Thus, the value of C_{gs} is proportional to both gate length and gate width. Figure 1.31 presents measured gate-source capacitance for both the 1×300- and 0.5×300-μm device used to produce previous comparison figures in this section. Note that although the 1-μm gate length device has double the gate area of the 0.5-μm gate length device, the value of C_{gs} is not doubled. The depletion region is composed of charge located both directly underneath the gate electrode as well as fringing charge in the gate-drain and gate-source spacing (see Figure 1.20). This fringing charge does not scale directly with gate length, but remains nearly constant. Thus, gate-source capacitance does not scale in a strictly linearly manner with gate length.

Gate-source capacitance is also affected by channel doping density. Higher values of doping density cause higher values of C_{gs} to be observed. In practice, device technology advances that have allowed the production of shorter gate length

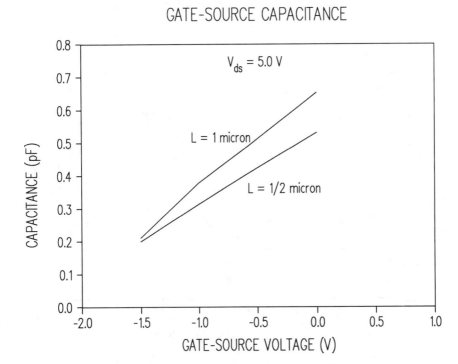

Figure 1.31 A comparison of measured microwave gate-source capacitance of MESFETs with 1- and 0.5-μm gate lengths.

devices have also required higher doping densities in the channel for optimum scaling. While shorter gate lengths produce smaller C_{gs} values, higher doping densities result in higher capacitance. These two trends have tended to cancel each other, resulting in a gate-source capacitance to gate width ratio, C_{gs}/Z, for microwave devices that has remained nearly constant at a value near 1 pF/mm.

The gate-drain capacitance of the MESFET is clearly closely related to the gate-source capacitance. In the normal mode of operation for amplifiers and oscillators, the primary characteristic affected by C_{gd} is the reverse isolation of the device. The smaller the gate-drain capacitance, the greater the isolation.

Figure 1.32 presents measured microwave gate-drain capacitance as a function of gate-drain voltage. The capacitance is seen to increase monotonically with increasing gate-drain voltage. The bias levels presented in the figure represent the range of bias levels for typical amplifier, oscillator, and frequency converter applications.

GATE-DRAIN CAPACITANCE

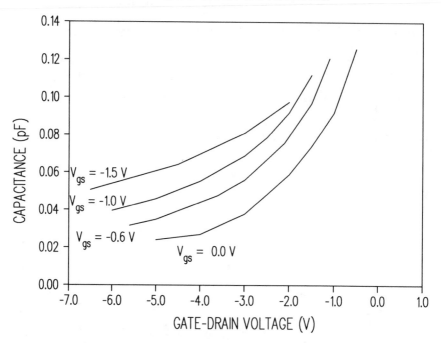

Figure 1.32 Measured microwave gate-drain capacitance of a 0.5×300-μm MESFET as a function of gate-drain voltage.

As in the case of gate-source capacitance, the value of C_{gd} is directly proportional to gate width. A similar relationship to gate length and doping density is also observed.

1.4 SECOND-ORDER EFFECTS

Many of the basic operating principles of MESFETs were presented in the previous section. Certain physical phenomena, however, cause observed MESFET behavior to deviate from this basic description under particular operating conditions. For some applications, an understanding of these differences between first-order theory and actual performance can be critical to the achievement of success in a circuit or device design. In particular, the low-frequency dispersion of device characteristics, the behavior of the device near pinch-off, and the performance of the device under conditions of optical stimulation often affect performance predictions in first-order ways.

1.4.1 Output Resistance as a Function of Frequency

Many of the electrical characteristics of GaAs MESFETs shift dramatically in value at extremely low frequencies (below about 1 MHz). A shift in the value of output resistance [14–22], transconductance [14, 18, 19, 23, 24], and device capacitance [18] can be observed. As frequency is increased above dc, measured device output resistance can drop by as much as an order of magnitude. Figure 1.33 shows typical measured output resistance as a function of frequency from 1 Hz to 1 MHz. The device characteristics illustrated in the figure show a transition from high values of output resistance to lower values at approximately 100 to 1000 Hz. Although the transition from high to low output resistance is typical of GaAs MESFETs, the frequency of the transition can vary dramatically from below 10 Hz to more than 1 MHz. Very similar frequency dispersion characteristics have also been observed in HEMTs [25].

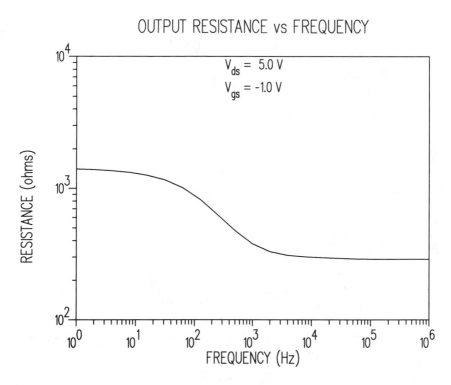

Figure 1.33 Typical output resistance as a function of frequency for a $0.5 \times 300\text{-}\mu m$ gate length MESFET.

The low-frequency dispersions of device characteristics are important to understand because of the implications for large-signal device modeling. Although the frequency dispersion effects take place well below the frequency bands of typical interest to the microwave engineer, they still play a role in the large-signal properties of the device. To produce accurate performance predictions, large-signal device models must describe both the dc and ac characteristics of the device. The existence of low-frequency dispersion effects also has significant implications for the parameter extraction processes described in Chapter 4.

Dispersion characteristics of MESFETs and HEMTs can vary significantly from device to device. This is especially true for devices fabricated using different processes or at different facilities. The starting semiconductor material quality and details of each fabrication processing step as well as the type of passivation applied to the surface affect dispersion properties. In this section, characteristics are presented for one particular device that is typical of GaAs MESFETs.

The output resistance of a MESFET as a function of frequency is shown in Figures 1.34 and 1.35. The MESFET device has a pinch-off voltage of about -2.0

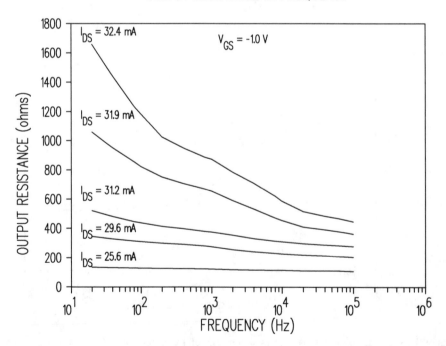

Figure 1.34 Measured output resistance as a function of frequency for a 0.8×400-μm MESFET. The frequency range is 20 Hz to 100 kHz and the gate-source voltage level is -1.0 V.

OUTPUT RESISTANCE vs FREQUENCY

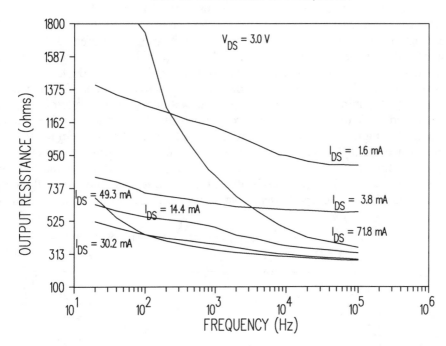

Figure 1.35 Measured output resistance as a function of frequency for a 0.8×400-μm MESFET. The frequency range is 20 Hz to 100 kHz and the drain-source voltage level is 3.0 V.

V and a maximum saturated drain current, I_{dss}, of approximately 70 mA. For the measurements presented in Figure 1.34, the device is biased with a gate-source voltage of $V_{gs} = -1.0$ V. This corresponds to the depletion region extending approximately midway into the channel. The drain-source current is varied from 25.6 to 32.4 mA as indicated on the curves. This corresponds to drain-source voltage variations from $V_{ds} = 1.0$ to 5.0 V in 1-V steps. For low drain-source bias levels, the observed shift in output resistance is insignificant. The amount of observed shift, however, increases monotonically with increased drain-source voltage. Figure 1.35 presents measured data from the same device biased at $V_{ds} = 3.0$ V. The percentage shift in output resistance is greatest for high drain-source current levels. These bias levels correspond to a very shallow depletion region under the gate (i.e., located near the semiconductor surface). As the reverse gate bias is increased, the depletion region is forced to extend deeper in the device, the channel current is decreased, and the observed dispersion in RF output resistance is decreased.

The observed characteristic frequency of the shift in resistance is also important. For low gate-source bias values (i.e., high currents), the characteristic frequency is below 20 Hz. As the device is biased nearer pinch-off, however, a second, less pronounced shift in resistance is also observed. This second shift occurs at a much higher frequency (closer to 1 kHz). Referring to Figure 1.34, we can see that a less significant higher frequency shift in resistance is also observed for drain-source bias levels greater than about 4 or 5 V (i.e., $I_{ds} > 31.9$ mA). The occurrence of this higher frequency shift is at bias levels that correspond to a depletion region at the drain end of the gate extending deep into the active channel of the device.

The 1-kHz shift becomes more pronounced at lower operating temperatures, as illustrated in Figure 1.36 where the measured output resistance as a function of frequency is plotted for three different temperatures. As the temperature is decreased from 325 to 275 K, the dominant low-frequency shift moves lower in frequency while the 1-kHz shift becomes more conspicuous. Although the 300 and 275 K curves are very close to each other, they do not cross. To obtain the data presented in Figure 1.36, the device is biased very near pinch-off. When the device

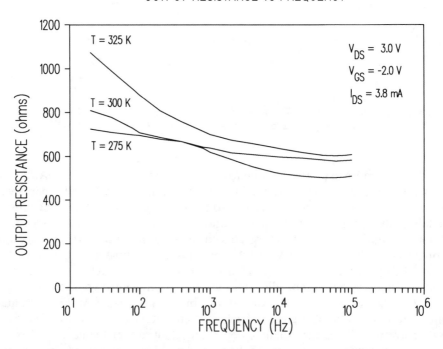

Figure 1.36 Measured output resistance as a function of frequency for various temperatures. The frequency range is 20 Hz to 100 kHz and the gate-source voltage level is near pinch-off. Although the curves corresponding to 300 and 275 K approach each other at approximately 800 Hz, the curves do not cross.

is biased well away from pinch-off, the 1-kHz output resistance shift is not observed even at 275 K (as shown in Figure 1.37). For this case, the dominant low-frequency shift in output resistance is again observed to move to lower frequencies as the temperature is reduced from 325 to 275 K.

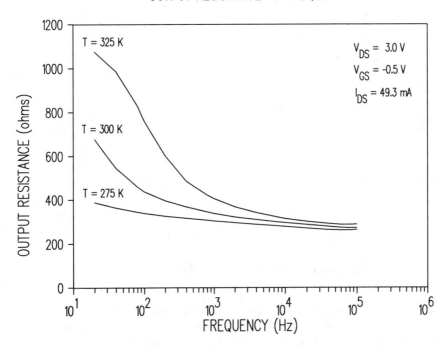

OUTPUT RESISTANCE vs FREQUENCY

Figure 1.37 Measured output resistance as a function of frequency for various temperatures. The frequency range is 20 Hz to 100 kHz and the gate-source voltage level is -0.5 V.

1.4.2 Transconductance as a Function of Frequency

Low-frequency shifts in transconductance values are also observed for microwave MESFETs and HEMTs. The shifts are typically much smaller than shifts in output resistance. Shifts on the order of 5 to 25% of the dc value are common. Because modeled transconductance values affect gain predictions in a first-order manner, these small drops in transconductance are very important for device modeling applications.

Transconductance dispersion for the same device used to produce the data of Figures 1.34 through 1.37 is presented in Figure 1.38. The observed shift in this

RF TRANSCONDUCTANCE

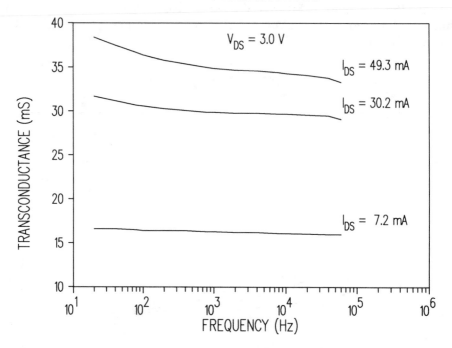

Figure 1.38 Measured transconductance as a function of frequency for a 0.8×400-μm MESFET.

value is seen to be much smaller than the corresponding shift in output resistance. The dc to microwave drop in transconductance for this particular device is found to be approximately 16% of the dc value for a wide range of bias values.

1.4.3 Subthreshold Effects

Although subthreshold current conduction is not observed under many operating conditions, it can be important for some applications. The physical phenomena that dominate device performance when a MESFET is biased near pinch-off are different than those that contribute most to behavior under normal operating conditions. As a result, channel pinch-off occurs more gradually than predicted by most models. This phenomena can be extremely important for digital applications and for certain mixer topologies.

The threshold voltage is defined theoretically as the applied gate voltage for which the channel is completely depleted of free carriers. The definition assumes that the expulsion of free carriers within the depletion region is total and that the substrate is a perfect insulator. Although both of these assumptions are good approximations over most of the operating range of the device, neither is absolutely true.

In an actual device, the boundary between the edge of the depletion region and the undepleted channel is gradual and occurs over a distance of several Debye lengths [equation (1.9)]. A graded transition also occurs between the active channel and the substrate. As the device is pinched off, these two graded regions are forced closer to each other.

Figure 1.39 illustrates the free carrier distribution in the channel directly below the gate electrode for two gate bias levels. The carrier density is plotted along the y-axis while the depth into the device channel is plotted along the x-axis. The

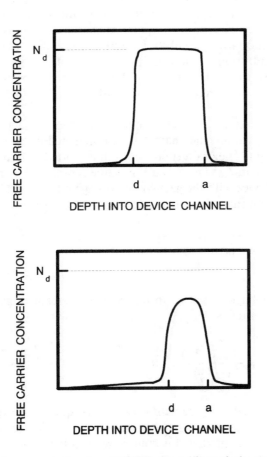

Figure 1.39 Free carrier concentration in a MESFET channel beneath the gate: (a) for a bias range typical of amplifier operation; (b) for a bias near pinch-off.

origin on the x-axis corresponds to the gate electrode and the point labeled "a" corresponds to the channel-substrate interface. In Figure 1.39(a), the two transition regions are separated slightly. The reverse bias on the gate causes the channel free carrier concentration to be nearly zero close to the gate. At the point labeled "d" the channel carrier concentration is in transition from totally depleted to totally undepleted. In the undepleted channel, the free carrier charge is approximately equal to the background donor density. Finally, near the channel-substrate interface, the free carrier concentration is again in transition.

As more reverse bias is applied to the gate, the transition region between the depleted and undepleted channel begins to meet the transition region between the channel-substrate interface. This situation is depicted in Figure 1.39(b). Under these conditions, the free carrier density in the channel never reaches the background donor density level. No portion of the channel is completely undepleted. In addition, the high electric field values at the channel-substrate interface can cause carrier injection into the substrate. Both of these effects cause the total channel carrier concentration to decline less rapidly with applied reverse gate bias than predicted by the simple theory. Thus, channel current near pinch-off is slightly higher than predicted by most current models and pinch-off occurs more gradually.

Figure 1.40 presents measured current-voltage characteristics of a GaAs MESFET. Also shown are the characteristics predicted by a standard square law modeling expression. The drain-source current is plotted in terms of the percent of the saturation current, I_{dss}. The I_{dss} current value is obtained by measuring the current level of the device with the gate-source voltage set to zero ($V_{gs} = 0$ V) and the drain-source current in saturation (typically, $V_{ds} > 2.5$ V).

$$(\%I_{dss}) = \frac{I_{ds}}{I_{dss}} \times 100 \qquad (1.25)$$

The gate-source voltage is plotted in terms of the percent of gate voltage applied between pinch-off and 0 V.

$$(\%V_{gs} - V_T) = \frac{(V_{gs} - V_T)}{(-V_T)} \times 100 \qquad (1.26)$$

The figure shows excellent agreement between the square law expression and measured results over most of the gate bias range. As the device is biased very near pinch-off, however, the predicted current is seen to fall rapidly toward zero while the measured current does not.

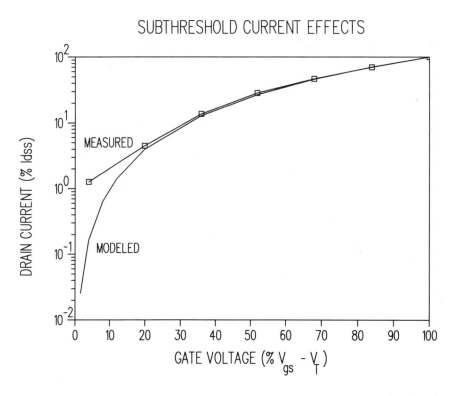

Figure 1.40 A comparison between measured and modeled drain-source current as a function of gate-source voltage. The curves are shown in a semilog plot to illustrate the failure of many models to describe subthreshold effects.

1.4.4 Optical Effects

When photons of sufficient energy strike a semiconductor, electron-hole pairs are produced in the material. These carriers contribute to the charge, electric field, and current distributions within the semiconductor. When the semiconductor material being illuminated is a MESFET or HEMT, the characteristics of the device are altered significantly. This property can be exploited to advantage for such applications as switching, injection locking of microwave oscillators, mixing, or photodetection. Advantages offered by photonic systems of this sort include high isolation between circuits and low interference from other electromagnetic signals.

If electrical systems are to exploit these optical-semiconductor interactions, accurate characterization of the behavior is required. The amount and type of characterization, however, are vast. Photon interaction with semiconductor material affects both dc and ac properties of the device. Modulated optical stimulation affects the device in a different manner than unmodulated stimulations. In addition, both the electrical signal and optical signal intensities must be considered. Although some models have been developed [25–28] to describe certain device-photon behavior, none has been successful at describing the entire range of behavior.

When a MESFET is illuminated suddenly by light of an appropriate wavelength, both a slow electrical response to the optical stimulation (taking on the order of several milliseconds to seconds) and a fast response (apparently instantaneous) occur. The slow response affects only the properties of a FET exposed to unmodulated light (or at modulation frequencies below about 10 kHz). The fast response, in contrast, contributes to the performance under both unmodulated and modulated light conditions.

The primary effect of unmodulated light on the dc characteristics of the device is to increase the current through the device. This increase in current is most apparent near device pinch-off. Figure 1.41 presents the threshold voltage as a function

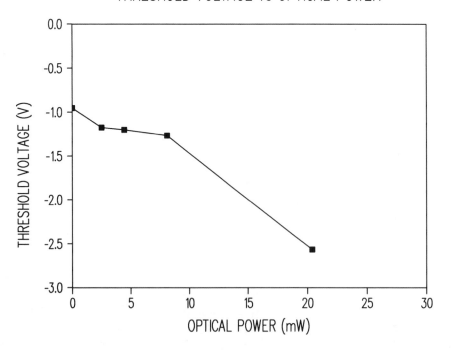

Figure 1.41 Threshold voltage as a function of optical power.

of optical power from the laser for a commercially available 0.5×200-μm gate length MESFET. The threshold voltage for this device when measured in the absence of optical illumination is $V_T = -0.95$ V. As the optical power is increased, threshold voltage is shifted to increasingly negative voltage values. The dc output resistance and dc transconductance are also altered under conditions of unmodulated illumination. Transconductance is found to increase under illuminated conditions for bias levels near device pinch-off, but to decrease for higher gate biases. Table 1.3 presents measured dc characteristics of a 0.5×200-μm gate length GaAs MESFET under conditions of no illumination and unmodulated laser illumination.

Table 1.3
Device Characteristics under Conditions of No Illumination and Unmodulated Illumination*

Measured dc Device Characteristic	Device Characteristic Value
I_{ds} dark	42.0 mA
I_{ds} unmodulated light	48.3 mA
G_m dark	33.1 mS
G_m unmodulated light	34.2 mS
R_{ds} dark	10000 Ω
R_{ds} unmodulated light	1683 Ω

*The optical output power for the laser is 4.4 mW at a wavelength of 840 nm. The dc characteristics were measured under identical voltage bias conditions.

The RF characteristics of MESFETs also shift when the device is exposed to unmodulated illumination. Output resistance and input capacitance increase under these conditions [29]. Transconductance also changes. Figure 1.42 presents measured output resistance as a function of frequency for three illumination conditions: dark, unmodulated, laser illumination, and modulated laser illumination. As in the case of the dc characteristics, RF transconductance increases for some bias conditions and decreases for others. The shifts in transconductance are also less significant than for output resistance.

Characteristics of the MESFET under conditions of only RF modulated light stimulation are qualitatively similar to the characteristics observed for unmodulated stimulation. Modulated laser stimulation increases dc current in the device by larger amounts, however, than unmodulated stimulations for the same laser drive level. This is shown in Figure 1.43 in which drain-source current is plotted

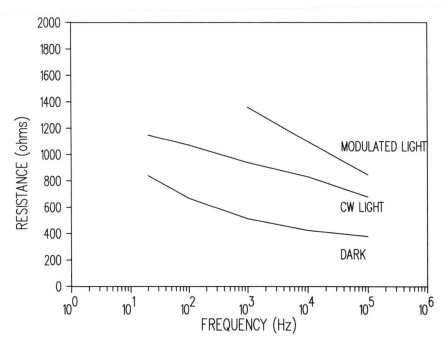

Figure 1.42 Output resistance as a function of frequency for dark conditions and conditions of unmodulated and modulated illumination.

against gate-source voltage. The optical power incident on the device is identical for both the modulated and unmodulated illumination cases. The modulation frequency used for these experiments is 100 kHz.

The RF characteristics of the MESFET under modulated illumination also show qualitative agreement with characteristics obtained using unmodulated stimulation. Note that measurement RF of the properties of MESFETs under conditions of modulated stimulation requires that no ac electrical stimulation be applied. Alternating voltages and currents are induced in the device from the modulated illumination and measured across external resistors. Both the transconductance and output resistance shifts observed for these conditions are much greater than those observed for unmodulated light of the same intensity. These data are summarized in Table 1.4. The RF output resistance under modulated light conditions is also plotted in Figure 1.42.

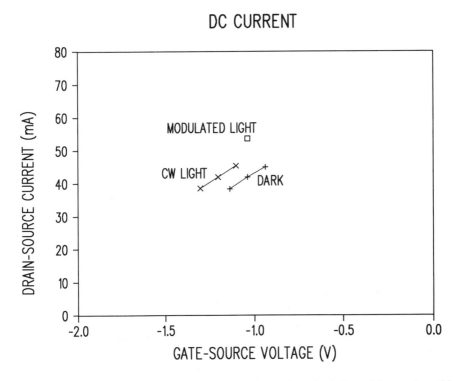

Figure 1.43 Drain-source current as a function of gate-source voltage for dark conditions and conditions of unmodulated and modulated illumination.

Table 1.4

Measured Shifts in Device Characteristics in the Presence of CW and Modulated Laser Stimulation at a Wavelength of 840 nm*

Device Characteristic	Dark	Shift under CW Illumination (%)	Shift under Modulated Illumination (%)
dc I_{ds}	42.0 mA	15	28
RF g_m	31.0 mS	17	28
RF r_{ds}	380 Ω	77	122

*All shifts are toward higher values. The 0.5×200-μm MESFET characteristics are: drain-source voltage, $V_{ds} = 3.0$ V; laser power level, $P_L = 4.4$ mW; modulation frequency, $f_{mod} = 100$ kHz.

1.5 HEMT DESCRIPTION

The high electron mobility transistor is a heterostructure field effect device. The term "high electron mobility transistor" is applied to the device because the structure takes advantage of the superior transport properties (high mobility and velocity) of electrons in a potential well of lightly doped semiconductor material. Other names commonly applied to the device include *two-dimensional electron gas field effect transistor* (TEGFET), *heterostructure FET* (HFET), *selectively doped heterostructure transistor* (SDHT), and *modulation-doped FET* (MODFET). Each of these names refers to some aspect of the device operation.

Figure 1.44 presents a cross-sectional view of a conventional HEMT structure. As in the MESFET, three metal electrode contacts (source, gate, and drain) are made to the surface of the semiconductor structure. The source and drain contacts are ohmic while the gate is a Schottky barrier. Further similarities between the HEMT and MESFET diminish rapidly as the description of operation proceeds to the important phenomena taking place within the semiconductor structure. A quick comparison between Figure 1.44 of the HEMT and Figure 1.20 of the MESFET shows the HEMT structure to be significantly more complex. This complexity is associated with fabrication difficulties, added costs, and lower yields. The primary motivations for pursuing such a structure are significant improvements in the device noise figure as well as some improvements in high-frequency performance.

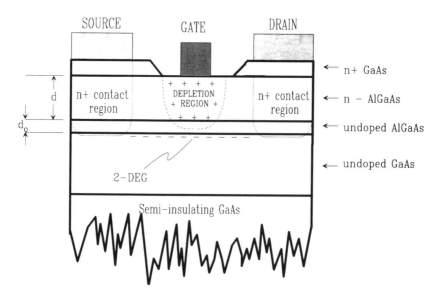

Figure 1.44 A conventional HEMT structure.

In practice, other types of layered semiconductor structures are often used in the fabrication of microwave and millimeter wave HEMTs. Figure 1.45 presents three alternative layered structures that have been employed in the realization of HEMT devices [30–32]. Figure 1.45(a) illustrates the layers required in the fabrication of a pseudomorphic HEMT. Note that this structure is similar to the layers illustrated in the conventional HEMT of Figure 1.44, but utilizes an additional layer of undoped InGaAs. HEMTs based on InP materials technology also offer certain attractive features. Figure 1.45(b) illustrates the material layers required to achieve HEMTs using this technology. Finally, Figure 1.45(c) illustrates the layers used in a multiple HEMT device. Such devices have the potential to increase the power handling capability of HEMTs. As in the MESFET, doping within the doped semiconductor layers is sometimes varied as a function of depth into the device.

Important geometric dimensions of the HEMT include the surface geometry dimensions (L, Z, L_{gs}, L_{gd}, L_s, and L_d), which are illustrated for the MESFET in Figure 1.21. The physical layouts of HEMT terminals are identical to MESFET terminal layouts, so that Figures 1.22 and 1.23 apply to both HEMT and MESFET devices. As with the MESFET, the most important dimension characterizing the HEMT physical structure is the gate length, which is labeled "L" in Figure

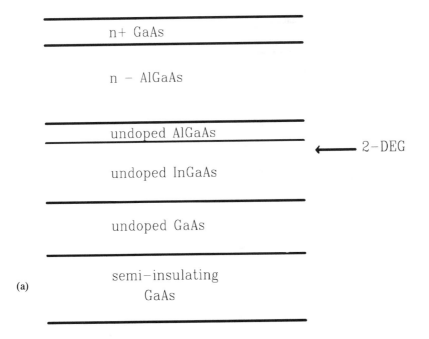

Figure 1.45 Alternative HEMT structures: (a) pseudomorphic HEMT; (b) pseudomorphic HEMT based on InP technology; (c) multiple quantum-well HEMT.

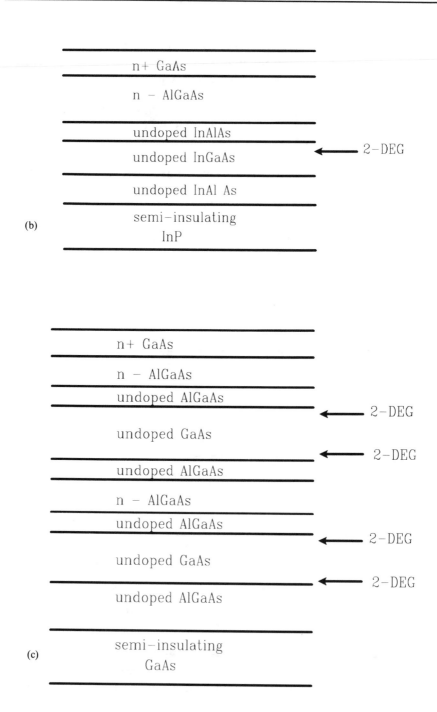

Figure 1.45 continued

1.21. This dimension is critical in determining the maximum frequency limits for HEMTs. Gate width and other surface geometry dimensions affect HEMT performance in a manner similar to the way these dimensions affect GaAs MESFET performance. Likewise, the range of gate length and gate width values used in the fabrication of HEMTs is typically identical to that used in MESFET fabrication. Each of these issues was discussed in Section 1.3.

HEMT structures differ significantly from MESFET structures in terms of the HEMT semiconductor layers. For the HEMT structure of Figure 1.44, the thickness of both the n-type AlGaAs and the undoped AlGaAs spacer layer are critical in determining device behavior. These layers are labeled "d" and "d_0," respectively, in the figure. The n-type AlGaAs layer thickness is typically on the order of 0.03 to 0.2 μm while the spacer layer thickness is typically about 50 Å (0.005 μm). For the more complex structures of Figure 1.45, additional vertical layer dimensions become important for modeling applications.

1.5.1 Current-Voltage Characteristics

The cross section of a simplified HEMT device structure is illustrated in Figure 1.46. This structure will serve as the focus of our discussion of the first-order operating principles of the HEMT. In the figure, a wide-band-gap semiconductor material (doped n-type) lies on an undoped narrow-band-gap material. The most common materials used for this structure are AlGaAs and GaAs, respectively, as shown in the figure. The thickness and doping density of the AlGaAs layer are designed so that this layer is completely depleted of free electrons under normal operating conditions. The undoped GaAs thickness is much less critical to the device design. However, the GaAs material quality (i.e., absence of defects and deep levels, *et cetera*) must be high.

The band diagram for this structure under conditions of zero gate bias is illustrated in Figure 1.47. A sharp dip in the conduction band edge occurs in the HEMT at the AlGaAs/GaAs boundary. As discussed in Section 1.2.4, this results in a high carrier concentration in a narrow region along the GaAs side of the heterojunction. The high free-electron concentration occurs over such a thin region that it is described as a two-dimensional electron gas (2-DEG) and quantified in terms of a sheet carrier density, n_s. Electrons traveling in this region do not encounter ionized donor atoms because the GaAs is undoped. Referring to Section 1.1, and particularly to Figure 1.5, we see that electron mobility is highest for lightly doped material. Thus, transport properties in the 2-DEG are favorable for fast response times and high-frequency operation. As with the MESFET, both depletion mode and enhancement mode devices can be fabricated.

Contact with the 2-DEG is made via the heavily doped, low-resistance source and drain wells. For low values of drain-to-source bias, a current flows from drain to source through the electron gas. In depletion mode devices, current will flow in

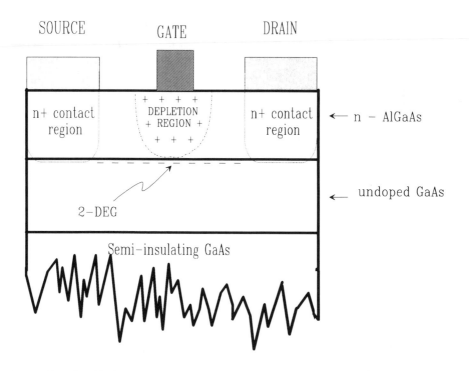

Figure 1.46 A simplified HEMT structure.

the absence of applied gate bias. This current is proportional to the applied drain-source voltage. As drain-source bias levels are increased, electron velocity and the current levels saturate. The saturated current level is determined primarily by the sheet carrier density of 2-DEG that forms in the structure.

The 2-DEG density is controlled by the gate bias. Increasing the negative bias applied to the gate decreases the depth (in electron energy) of the potential well at the AlGaAs/GaAs boundary. For depletion mode devices, this results in a diminished sheet carrier density of the electron gas and, therefore, reduced current conduction. For enhancement mode devices, no current conduction takes place until forward gate bias is applied. Note that in the MESFET, bias on the gate terminal controls the depth of the undepleted channel, while in the HEMT, gate bias controls the carrier density. Both of these effects, however, result in control of the maximum channel current. For large enough values of reverse gate bias, the sheet carrier concentration becomes negligibly small and the channel current is pinched off. As in the case of the MESFET, the gate voltage level that corresponds to this phenomena is termed pinch-off voltage.

The current-voltage relationships expected from an ideal HEMT as described above are very similar to the ideal MESFET characteristics of Figure 1.24(a). The

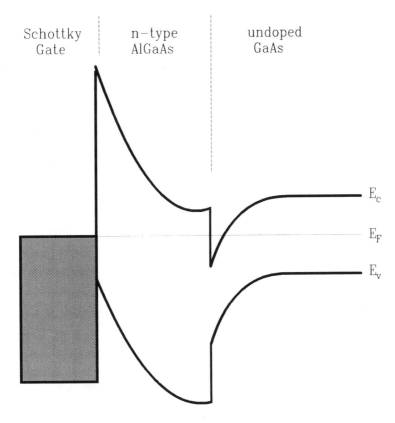

Figure 1.47 Band diagram corresponding to the simplified HEMT structure of Figure 1.46.

characteristics of actual HEMT devices are similar to these ideal characteristics, with some important exceptions. As in the case of the MESFET, actual I_{ds}-V_{ds} characteristics exhibit finite slope even after semiconductor-limited velocity has been reached. A more important distinction occurs for forward gate bias conditions or at low values of reverse gate bias. Note that if the bias applied to the gate is not sufficient to deplete the AlGaAs material completely, then a conduction path between the source and drain electrode is opened up through this layer. Conduction behavior through this layer is similar to conduction in a MESFET—in this case, a MESFET fabricated from highly doped AlGaAs. Electron transport properties in this parasitic MESFET are significantly inferior to the properties in the 2-DEG. Thus, device performance, in terms of current drive and transconductance, begins to degrade when parasitic MESFET conduction occurs.

The conventional HEMT structure of Figure 1.44 shows two important semiconductor layers not described in the preceding first-order explanation or shown in the simplified structure of Figure 1.46. The more complete structure has a highly doped GaAs layer present on the surface of the semiconductor. All of the structures shown in Figure 1.45 utilize a similar layer. This layer serves to reduce the ohmic resistance of the source and drain contacts. Typically, the layer is etched away where the gate contact is to be formed. A second distinction between Figures 1.44 and 1.46 is the existence of an undoped AlGaAs layer between the *n*-type AlGaAs and undoped GaAs of Figure 1.44. This layer is referred to as a *spacer layer* and serves to separate the electrons flowing in the 2-DEG from the dopant ions in the wide-band-gap material. Without such a layer, electrons in the 2-DEG are scattered as they pass close to the ionized donors. This scattering reduces electron mobility and, therefore, diminishes the effect exploited in this device. Each of the structures illustrated in Figure 1.45 also utilizes an undoped spacer layer for this purpose. In the case of the multiple HEMT structure [Figure 1.45(c)], several spacer layers are utilized.

Figure 1.48 presents the I_{ds}-V_{ds} characteristics measured for a 0.7×200-μm pseudomorphic HEMT. The characteristics are seen to be very similar qualitatively to the MESFET characteristics of Figure 1.24(b). As in the case of the MESFET, the magnitude of the drain-source current of a given structure is directly proportional to device gate width. Gate length also plays a role in HEMT current behavior similar to the role in a MESFET. Reduced gate length on identical semiconductor layers increases saturation current levels and output conductance levels slightly. Figure 1.48(b) presents measured drain-source current as a function of gate-source voltage. Although these characteristics share similarities with the corresponding MESFET characteristics of Figure 1.24(c), a significant difference is apparent. In the HEMT characteristics, the rate of increase in drain-source current with reduced reverse gate-source bias is seen to decline significantly at a bias level of approximately -0.3 V. This corresponds to the onset of conduction via the parasitic MESFET within the HEMT structure. This phenomenon also affects device transconductance as discussed in Section 1.5.2.

1.5.2 Transconductance and Output Resistance

The definitions of device output conductance and transconductance are given by equations (1.20) and (1.21). These definitions apply not only to the MESFET, but also to the HEMT or any other two-port device. The definitions focus attention on the derivatives of the curves presented in Figure 1.48.

As in the case of the MESFET, high values for output resistance (i.e., the inverse of output conductance) and transconductance are desired to obtain high-gain and high-frequency performance. The HEMT structure takes advantage of superior electron transport properties to produce high values of transconductance. Likewise, the carrier confinement obtained in the 2-DEG of a HEMT structure contributes to high output resistance values.

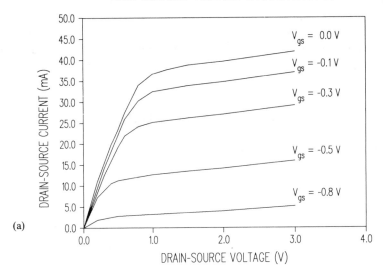

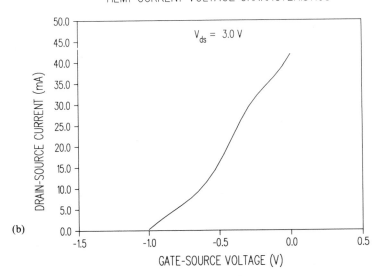

Figure 1.48 Measured current-voltage characteristics for a 0.7 × 200-μm pseudomorphic HEMT: (a) drain-source current as a function of drain-source voltage; (b) drain-source current as a function of gate-source voltage.

Figure 1.49 presents measured microwave output resistance for a 0.7×200-μm HEMT as a function of drain-source voltage. The curves are as expected from the definition of output resistance and from examination of Figure 1.48(a). The curves are also seen to be very similar to the corresponding MESFET characteristics of Figure 1.25.

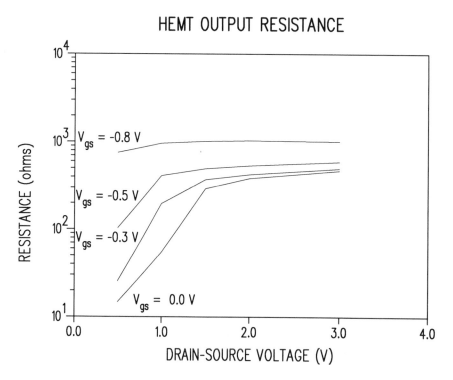

Figure 1.49 Measured microwave output resistance as a function of drain-source voltage for a 0.7×200-μm pseudomorphic HEMT.

Both device dimensions and material properties affect the output resistance of the HEMT. As in the MESFET, output resistance is inversely proportional to device gate width. Reduced gate length also results in moderately reduced output resistance. The doping density or doping density profile in the n-type AlGaAs layer along with the dimension of the undoped AlGaAs spacer layer also contribute to the magnitude of the device output resistance. In general, lower doping levels and thicker spacer layers tend to increase output resistance. Unfortunately, these characteristics also lead to reduced transconductance.

The existence of deep levels within the HEMT structure—and particularly at the GaAs/AlGaAs boundary—plays an important role in determining the output resistance of HEMTs. As in the MESFET, output resistance of HEMTs is observed to decrease dramatically between dc and a frequency of around 1 MHz [25]. The HEMT also exhibits shifts in electrical characteristics under conditions of optical stimulation that are similar to those observed for MESFETs [33].

Figure 1.50 presents measured microwave transconductance of a 0.7×200-μm gate length device as a function of gate-source voltage. These curves are strikingly dissimilar to the corresponding MESFET curves of Figure 1.27. The transconductance is seen to be low near device pinch-off; it begins to increase as reverse gate bias is reduced. This portion of the curves shows some similarity to MESFET

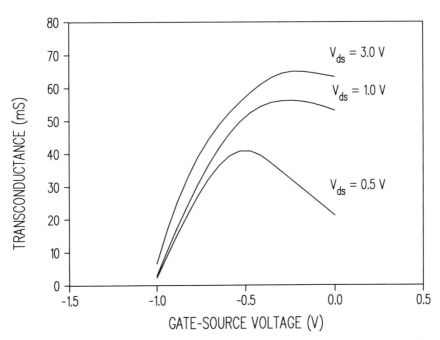

Figure 1.50 Measured microwave transconductance as a function of drain-source voltage for a 0.7×200-μm pseudomorphic HEMT.

behavior. In contrast to the MESFET, however, the HEMT transconductance begins to decrease rapidly at some gate bias level. As explained previously, this decrease in transconductance is the result of the onset of conduction through the parasitic MESFET in the AlGaAs layer. Note that for the $V_{ds} = 3.0$ V curve of Figure 1.50, the gate voltage corresponding to peak transconductance is approximately -0.3 V. As expected, this corresponds to the gate voltage where the rate of increase of drain-source current was observed to fall off in Figure 1.48(b). The bias conditions at which parasitic conduction begins to contribute to device characteristics are determined by the doping densities and dimensions of the AlGaAs layers. Normally, this operating region of degraded transconductance performance is avoided.

The magnitude of device transconductance is directly proportional to gate width and approximately inversely proportional to gate length. As in the case of the MESFET, gate length dependence is not strictly linear because of second-order effects. Transconductance of the HEMT is also greatly affected by semiconductor layer dimensions and channel material properties as mentioned in the Section 1.5.1.

1.5.3 Capacitance-Voltage Characteristics

Because the terminal configurations of MESFETs and HEMTs are nearly identical, defining the same important device capacitances for the two devices is useful: gate-source capacitance and gate-drain capacitance. Definitions for these capacitances are presented in equations (1.23) and (1.24). Under normal HEMT operating conditions, the depletion region under the gate extends throughout the n-type AlGaAs layer. Therefore, the depth of the depletion region penetration into the semiconductor is fixed. This condition is violated only when the parasitic MESFET conduction begins to influence device behavior. Because this situation differs from the case of the MESFET, observed HEMT capacitance characteristics are somewhat different from those observed in MESFETs.

Figure 1.51 presents measured gate-source capacitance as a function of gate-source voltage for a 0.7×200-μm HEMT device. These characteristics are seen to be significantly different qualitatively than the corresponding MESFET characteristics of Figure 1.30. Near pinch-off, the capacitance is low. As the device is turned on, gate-source capacitance increases abruptly. Over most of the normal operating range of the device, however, little variation in capacitance is seen. This is consistent with the fact that depletion region depth is fixed. Also observable in the characteristics is a slight decline in capacitance corresponding to the onset of parasitic MESFET conduction.

As the gate dimensions of a device are altered, the area of the depletion region underneath the gate is also changed. This is identical to the situation for the

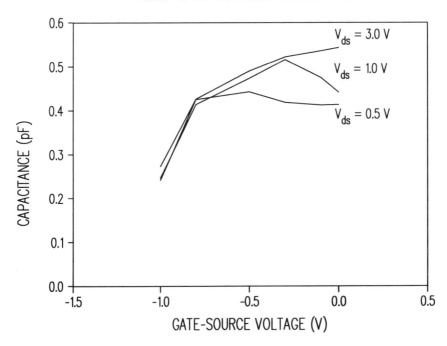

Figure 1.51 Measured gate-source capacitance as a function of gate-source voltage for a 0.7×200-μm pseudomorphic HEMT.

MESFET. Gate-source capacitance, therefore, is proportional to both gate length and gate width. Semiconductor layer thicknesses and doping profiles will also affect gate-source capacitance in first-order ways. In general, thinner layers and higher doping concentrations result in higher capacitance values.

Figure 1.52 presents measured microwave gate-drain capacitance as a function of gate-drain voltage. The measured capacitance value is obtained from microwave s-parameter data (0.045 to 20 GHz) as described in Chapter 4. Curves corresponding to two gate-source bias levels ($V_{gs} = 0$ V and $V_{gs} = -0.8$ V) are presented. Curves for all gate bias levels between these two voltages fall very close to the measured data presented. In contrast to the MESFET characteristics of Figure 1.32, the HEMT gate-drain capacitance is seen to be nearly constant over most of the operating range. Only at extremely low values of reverse gate-drain voltage does the HEMT gate-drain capacitance increase.

HEMT GATE-DRAIN CAPACITANCE

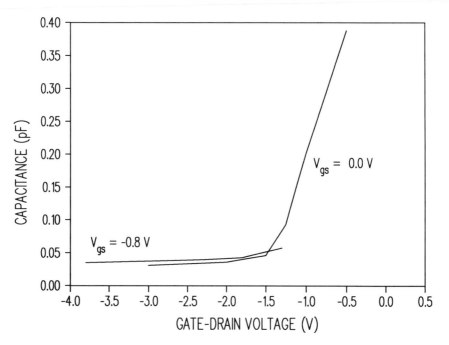

Figure 1.52 Measured gate-drain capacitance as a function of gate-drain voltage for a 0.7×200-μm pseudomorphic HEMT.

As in the case of gate-source capacitance, the value of C_{gd} is directly proportional to gate width. Observed relationships between C_{gd} and gate length or between C_{gd} and semiconductor layer thicknesses are similar to those determined for C_{gs}.

REFERENCES

[1]   R.H. Bube, *Electronic Properties of Crystalline Solids: An Introduction to Fundamentals,* New York: Academic Press, 1974.
[2]   S. Adachi, "GaAs, AlAs, and Al$_x$Ga$_{1-x}$As: Material Parameters for Use in Research and Device Applications," *J. Appl. Phys.,* Vol. 58, August 1, 1985, pp. 1–29.
[3]   R.F. Pierret, G.W. Neudeck, *Modular Series on Solid State Devices,* Vol. I, Reading, MA: Addison-Wesley, 1983.
[4]   S.M. Sze, *Physics of Semiconductor Devices,* New York: John Wiley and Sons, 1969.
[5]   R.S. Muller and T.I. Kamins, *Device Electronics for Integrated Circuits,* New York: John Wiley and Sons, 1977.

[6]   M.A. Littlejohn, T.H. Glisson, and J.R. Hauser, *GaInAsP Alloy Semiconductor,* New York: John Wiley and Sons, 1982.

[7]   J.M. Golio, R.J. Trew, G.N. Maracas, and H. Lefevre, "A Modeling Technique for Characterizing Ion-Implanted Material Using C-V and DLTS Data," *Solid-State Electron.,* Vol. 27, 1984, pp. 367–373.

[8]   M.A. Littlejohn, R.J. Trew, J.R. Hauser, and J.M. Golio, "Electron Transport in Planar-Doped Barrier Structures Using an Ensemble Monte Carlo Method," *J. Vac. Sci. Technol. B,* Vol. 1, April–June 1983, pp. 449–454.

[9]   D.P. Kennedy and R.R. O'Brien, "On the Measurement of Impurity Atom Distributions by the Differential Capacitance Technique," *IBM J. Res. Dev.,* Vol. 13, March 1969, pp. 212–214.

[10] D.P. Kennedy, P.C. Murley, and W. Kleinfelder, "On the Measurement of Impurity Atom Distributions in Silicon by the Differential Capacitance Technique," *IBM J. Res. Dev.,* Vol. 12, September 1968, pp. 399–409.

[11] W. Walukiewicz, L. Lagoaski, L. Jastrzebski, M. Lichtensteiger, and H.C. Gatos, "Electron Mobility and Free-Carrier Absorption in GaAs: Determination of the Compensation Ratio," *J. Appl. Phys.,* Vol. 50, February 1979, pp. 899–908.

[12] S.H. Wemple, W.C. Niehaus, H.M. Cox, J.V. DiLorenzo, and W.O. Schlosser, "Control of Gate-Drain Avalanche in GaAs MESFET's," *IEEE Trans. Electron Devices,* Vol. ED-27, June 1980, pp. 1013–1018.

[13] A.G. Milnes, D.L. Feucht, *Heterojunctions and Metal-Semiconductor Junctions,* New York: Academic Press, 1972.

[14] J.M. Golio, M. Miller, G.N. Maracas, and D.A. Johnson, "Frequency Dependent Electrical Characteristics of GaAs MESFETs," *IEEE Trans. Electron Devices,* Vol. ED-37, May 1990.

[15] M.A. Smith, T.S. Howard, K.J. Anderson, and A.M. Pavio, "RF Nonlinear Device Characterization Yields Improved Modeling Accuracy," *IEEE Microwave Theory Tech. Symp. Digest,* 1986, pp. 381–384.

[16] P. Canfield, J. Medinger, and L. Forbes, "Buried-Channel GaAs MESFET's with Frequency-Independent Output Conductance," *IEEE Electron Devices Lett.,* Vol. EDL-8, March 1987, pp. 88–89.

[17] L.E. Larson, "An Improved GaAs MESFET Equivalent Circuit Model for Analog Integrated Circuit Applications," *IEEE J. Solid-State Circuits,* Vol. SC-22, August 1987, pp. 567–574.

[18] P.H. Ladbrooke and S.R. Blight, "Low-Field Low-Frequency Dispersion of Transconductance in GaAs MESFET's with Implications for Other Rate-Dependent Anomalies," *IEEE Trans. Electron Devices,* Vol. ED-35, March 1988, pp. 257–267.

[19] N. Scheinberg, R. Bayruns, and R. Goyal, "A Low-Frequency GaAs MESFET Circuit Model," *IEEE J. Solid-State Circuits,* Vol. SC-23, April 1988, pp. 605–608.

[20] J.F. Wager and A.J. McCamant, "GaAs MESFET Interface Considerations," *IEEE Trans. Electron Devices,* Vol. ED-34, May 1987, pp. 1001–1004.

[21] C. Camacho-Peñalosa and C.S. Aitchison, "Modeling Frequency Dependence of Output Impedance of a Microwave MESFET at Low Frequencies," *Electron. Lett.,* Vol. 21, June 6, 1985, pp. 528–529.

[22] N. Ishihara, H. Kikuchi, and M. Ohara, "Gigahertz-Band High-Gain GaAs Monolithic Amplifiers Using Parallel Feedback Techniques," *IEEE J. Solid-State Circuits,* Vol. SC-24, August 1989, pp. 962–968.

[23] A. Zylberstejn, G. Bert, and G. Nuzillat, "Hole Traps and Their Effects in GaAs MESFETs," Chapter 1 in *Inst. Phys. Conf. Ser. No. 45,* 1979, pp. 315–325.

[24] S.R. Blight, R.H. Wallis, and H. Thomas, "Surface Influence on the Conductance DLTS Spectra of GaAs MESFETs," *IEEE Trans. Electron Devices,* Vol. ED-33, October 1986, pp. 1447–1453.

[25] J.A. Reynoso-Hernandez and J. Graffeuil, "Output Conductance Frequency Dispersion and Low-Frequency Noise in HEMTs and MESFETs," *IEEE Trans. Microwave Theory Tech.,* Vol. MTT-37, September 1989, pp. 1478–1481.

[26] D.A. Warren, J.M. Golio, and E. Johnson, "Simulation of Optically Injection Locked Oscillators Using a Novel SPICE Model," *IEEE Trans. Microwave Theory Tech.*, Vol. MTT-36, 1988, pp. 1535–1539.

[27] A.A. DeSalles, "Optical Control of GaAs MESFETs," *IEEE Trans. Microwave Theory Tech.*, Vol. MTT-31, October 1983, pp. 812–820.

[28] R. Soares, J. Graffeuil, and J. Obregon, eds., *Applications of GaAs MESFETs,* Norwood, MA: Artech House, 1983.

[29] J.L. Gautier, D. Pasquet, and P. Pouvil, "Optical Effects on the Static and Dynamic Characteristics of a GaAs MESFET," *IEEE Trans. Microwave Theory Tech.*, Vol. MTT-33, September 1985, pp. 819–822.

[30] P.M. Smith and A.W. Swanson, "HEMTs—Low Noise and Power Transistors for 1 to 100 GHz," *Appl. Microwave,* May 1989, pp. 63–73.

[31] U.K. Mishra, A.S. Brown, S.E. Rosenbaum, C.E. Hooper, M.W. Pierce, M.J. Delaney, S. Vaughn, and K. White, "Microwave Performance of AlInAs-GaInAs HEMTs with 0.2 and 0.1 Micron Gate Length," *IEEE Electron Devices Lett.*, Vol. EDL-9, December 1988, pp. 647–649.

[32] A.K. Gupta, J. Higgins, and C.P. Lee, "High Electron Mobility Transistors for Millimeter Wave and High Speed Digital Applications," *Characterization of Very High Speed Semiconductor Devices and Integrated Circuits,* SPIE Proc., Vol. 795, 1987, pp. 68–90.

[33] R.N. Simons, "Microwave Performance of an Optically Controlled AlGaAs/GaAs High Electron Mobility Transistor and GaAs MESFET," *IEEE Trans. Microwave Theory Tech.*, Vol. MTT-35, December 1987, pp. 1444–1455.

Chapter 2
DEVICE MODELS

A number of different semiconductor device models exist, and each model can be classified among several different categories. Models are often classified according to how they are derived. Empirical models are derived using an approach based on describing observed characteristics with arbitrary functions, while physically based models are derived from consideration of the physical principles that apply to the device structure. Many models have aspects of both empirical and physical derivations. Models are also often classified according to the type of calculations required to obtain predictions. Performance predictions are obtained from an analytical model by evaluating analytical mathematical expressions. In contrast, numerical models involve the use of numerical techniques to solve (often using iterative techniques) coupled differential equations and, hence, gain predictions of device behavior. The type of performance predicted by a model is also used as a classification scheme. Small-signal, large-signal, and noise models are utilized to obtain information concerning each of these kinds of device characteristics.

In a typical application, models are used to predict or estimate performance information that is not available or easily obtainable by direct measurement. Small-signal models can offer the designer the ability to predict performance of devices with gate width dimensions that have been scaled from previously measured devices. Another important use of these models is to interpolate or extrapolate measured data to frequencies not covered by measurements. Noise models are used to predict the noise figure for arbitrary circuit topologies, which incorporate a particular device, or to predict the ultimate noise performance of a device. A large-signal model provides a means of obtaining performance information concerning nonlinear operation of a device or device-circuit combination.

Both device designers and circuit designers find applications for device models. Because the lists of requirements that these two groups place on models are not necessarily identical, they often make use of different models. The distinctions between their respective requirements, however, diminish as circuit designers become more involved with design of integrated circuitry. Although few foundries today allow the circuit designer to direct arbitrary processing steps for devices, tasks such as design centering and yield analyses are leading the circuit designer to become familiar with the physically based models that had been used previously only by device designers.

This leads to a discussion of the merits of physically based models *versus* empirical models. Both physical and empirical modeling techniques are associated with certain advantages and disadvantages. The physically based models are very appealing to device designers. Such models are also useful to circuit designers who have some control over the fabrication process because they allow simultaneous optimization of both the devices and the circuits in which they are to be used. Additionally, a physical model is useful for predicting the effects of process variations on the electrical behavior of a device. Further, if the statistical distribution of the process parameters is known, yield predictions can also be obtained. Using such an approach, performance prediction information may be obtained purely from physical data describing the device (i.e., device geometry and semiconductor material properties). No electrical characterization of individual devices is required. The merits of this approach for device designers are obvious.

Unfortunately, purely physical models are not as accurate as required for most circuit design applications. The inaccuracies arise from the assumptions and approximations required to perform the device analysis. Even the most sophisticated multidimensional numerical device simulation techniques make simplifying assumptions that can severely limit their accuracy. For example, surface state effects, traps, nonhomogeneous interfaces, *et cetera,* are typically neglected in these analyses. Yet these phenomena have a significant effect on the microwave performance of modern devices. A second problem with physically based models is that information concerning the physical design of the device can often be difficult or impossible to obtain—especially for the circuit designer utilizing purchased devices. This problem may be solved by extracting the physical parameter values from measured data in the same way that empirical parameter values are determined.

In contrast, empirical models are capable of prediction accuracies that approach measurement capability. The primary difficulty with this approach is that large amounts of tedious characterization data are often required to obtain such accuracy. In addition, minor changes in the device geometry or material require the performance of complete recharacterizations. Such models are also of questionable value when performing design centering or yield analyses because the empirical parameter distributions do not vary independently.

Because physical models are typically less accurate than their empirical counterparts, one attractive implementation is to use an empirical model to simulate nominal device performance and then use the physically based model to predict the deviations about this nominal behavior resulting from process parameter variations.

The optimal model for any given application depends on many factors. The model requirements for particular applications will vary depending on the type of circuitry required and the point in the design and fabrication process at which the model is to be used. The availability of the model within a circuit simulation routine is clearly a key factor. Computational efficiency of the model and the accuracy of the model predictions are also important. No single model has emerged that optimally meets all current modeling needs. Instead, many different models are being used—often within the same circuit simulation package.

This chapter presents some details of several popular models. The chapter is divided into sections according to the type of performance predictions for which the models are used—small signal, large signal, and noise. Within each section, both physical and empirical approaches are discussed. Numerical device simulations are not addressed. The multidimensional numerical simulation approach has serious limitations when applied to circuit design problems. As a greater understanding of all of the device physics is gained and computer capability continues to expand, the role of these models in circuit simulation applications will increase.

2.1 SMALL-SIGNAL MODELS

The small-signal MESFET-HEMT model is extremely important for active microwave circuit work. These models provide a vital link between measured s-parameters and the electrical processes occurring within the device. Each of the elements in the equivalent circuit provides a lumped element approximation to some aspect of the device physics. A properly chosen topology, in addition to being physically meaningful, provides an excellent match to measured s-parameters over a very wide frequency range. When element values are properly extracted, the model is valid above the frequency range of the measurements, providing the possibility of extrapolating device performance to frequencies beyond some equipment's measurement capabilities. In addition, equivalent circuit element values can be scaled with gate width, thereby enabling the designer to predict the s-parameters of different size devices from a given foundry. The ability to include device gate width scaling as part of the circuit design process is important in MMIC design applications.

2.1.1 Physical Significance of Equivalent Circuit Element Values

A fairly standard MESFET-HEMT equivalent circuit topology is shown in Figure 2.1. Although other circuit topologies involving additional elements have been described in the literature, the topology of Figure 2.1 has been shown to provide an excellent match to measured s-parameters through 26 GHz. This topology has the additional advantage that the elements can be uniquely extracted (Chapter 4). In Figure 2.2, the same equivalent circuit is shown superimposed on a device cross section, indicating the physical origin of the equivalent circuit. A brief discussion follows of each equivalent circuit element and its role in modeling the device physics.

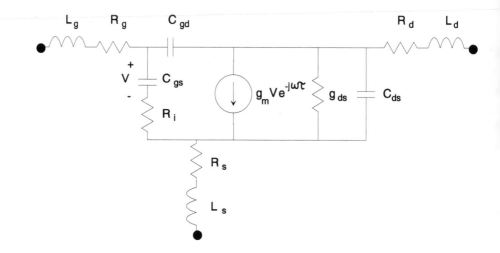

Figure 2.1 MESFET/HEMT small-signal model including parasitic elements.

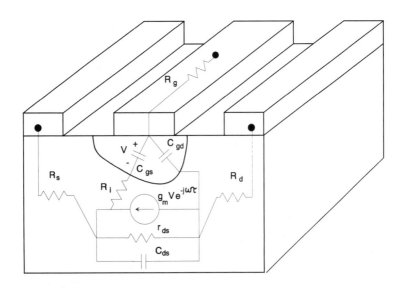

Figure 2.2 MESFET/HEMT small-signal model showing physical origin of elements.

2.1.1.1   Parasitic Inductances L_s, L_d, and L_g

The parasitic inductances arise primarily from metal contact pads deposited on the device surface. Because these values are dependent on the surface features of the device, they are essentially equal for MESFETs and HEMTs. For modern short gate length devices, the gate inductance is usually the largest of the three, although this is a function of the particular layout employed. Typically, L_g and L_d are on the order of 5 to 10 pH. The source inductance is often small, ~ 1 pH, especially for devices utilizing via hole grounds. Note that these inductances exist in addition to any parasitic bond wire inductances or parasitic package inductances, which must also be accounted for in the circuit model. In many cases, bonding inductances are on the order of 0.1 to 0.3 nH and dominate the device parasitics.

2.1.1.2   Parasitic Resistances R_s, R_d, and R_g

The resistances R_s and R_d are included to account for the contact resistance of the ohmic contacts as well as any bulk resistance leading up to the active channel. The gate resistance R_g results from the metalization resistance of the gate Schottky contact. All three resistances are on the order of 1 Ω for a modern microwave device. Also, R_s and R_d tend to be slightly less in HEMTs than in MESFETs. Although measurements indicate a slight bias dependence in these values, they are held constant in the large-signal models commonly available in circuit simulators today. These resistance values can be estimated either from forward conduction measurements or directly from the s-parameters using an optimization technique.

2.1.1.3   Capacitances C_{gs}, C_{gd}, and C_{ds}

As discussed in Chapter 1, the capacitances C_{gs} and C_{gd} model the change in the depletion charge with respect to the gate-source and gate-drain voltages respectively. Under typical amplifier or oscillator bias conditions, the gate-source capacitance is the larger quantity because it models the change in depletion charge resulting from fluctuations in gate-source voltage. Under these normal bias conditions, the gate-drain capacitance C_{gd} is considerably smaller in magnitude than C_{gs} but nevertheless critical to obtaining accurate s-parameter predictions. The drain-source capacitance C_{ds} is included in the equivalent circuit to account for geometric capacitance effects between the source and drain electrodes. It is usually not considered to be bias dependent for the purposes of device modeling. Values for C_{gs} are typically on the order of 1 pF/mm gate width under normal amplifier bias conditions. The values of C_{gd} and C_{ds} are about $\frac{1}{10}$ of the value of C_{gs}. Because of symmetry, C_{gs} and C_{gd} are approximately equal for $V_{ds} = 0$ and reverse roles for inverted drain-source bias conditions ($V_{ds} < 0$).

2.1.1.4 Transconductance g_m

The intrinsic gain mechanism of the FET is provided by the transconductance. The transconductance g_m is a measure of the incremental change in the output current I_{ds} for a given change in input voltage V_{gs}. Mathematically, it is defined as

$$g_m = \frac{\partial I_{ds}}{\partial V_{gs}} \tag{2.1}$$

As discussed in Section 1.4.2, the transconductance varies with frequency below a frequency of about 1 MHz. Transconductance values vary directly with gate width and inversely with gate length for both MESFETs and HEMTs. In practice, the observed transconductance of HEMTs is slightly higher than that of MESFETs with equivalent gate geometries.

2.1.1.5 Output Conductance g_{ds}

The output conductance is a measure of the incremental change in output current I_{ds} with the output voltage V_{ds}. The output conductance is defined mathematically as

$$g_{ds} = \frac{\partial I_{ds}}{\partial V_{ds}} \tag{2.2}$$

Values of g_{ds} are on the order of 1 mS/mm gate width at typical amplifier biases. Also, as gate length is reduced, output conductance tends to increase in both MESFETs and HEMTs. Even more significant than with the transconductance is the low frequency dispersion in the output conductance. The RF output conductance can be more than 100% higher than dc output conductance. As with the transconductance, the RF values are of primary concern for small-signal modeling applications.

2.1.1.6 Transconductance Delay

The transconductance cannot respond instantaneously to changes in gate voltage. The delay inherent to this process is described by the transconductance delay τ. Physically, the transconductance delay represents the time it takes for the charge to redistribute itself after a fluctuation of gate voltage. Typical values of τ are on the order of 1 pS for microwave MESFETs and HEMTs. From physical considerations, transconductance delays are expected to be shorter in HEMTs than in MESFETs with equivalent geometries and tend to decrease with decreasing gate length.

2.1.1.7 Charging Resistance

The charging resistance is included in the equivalent circuit primarily to improve the match to S_{11}. For many devices, however, the presence of R_g is sufficient to match the real part of S_{11}. In either case, R_i is difficult to extract and is of questionable physical significance. The inclusion of R_i also complicates the large-signal analysis.

2.1.2 Scaling Rules for Element Values

Scaling rules for the FET equivalent circuit are easily arrived at from simple analytical expressions for the elements or from measurements of different size devices from the same foundry. Most of the scaling rules are intuitively obvious and element values are easily scaled up or down based on the total gate width.

Because both drain-source current and depletion charge are proportional to total gate width Z, the partial derivatives of these quantities (i.e., g_m, g_{ds}, C_{gd}, and C_{gs}) are also expected to be proportional to Z. In addition, because C_{ds} is a geometric capacitance from source to drain, it is also proportional to gate width. The circuit parameters of some device with total gate width Z can be related to those of a device scaled to gate width Z' using the scale factor

$$s_1 = \frac{Z'}{Z}. \tag{2.3}$$

The above-mentioned five-element values are then easily scaled to some arbitrary gate width Z' as follows:

$$g'_m = g_m s_1 \tag{2.4}$$

$$g'_{ds} = g_{ds} s_1 \tag{2.5}$$

$$C'_{gd} = C_{gd} s_1 \tag{2.6}$$

$$C'_{gs} = C_{gs} s_1 \tag{2.7}$$

$$C'_{ds} = C_{ds} s_1 \tag{2.8}$$

Both the drain and source resistances are inversely proportional to gate width. This fact is easily deduced from Ohm's law since for either resistor the following expression can be written

$$R_{s,d} = V/I_{ds} \tag{2.9}$$

where V is the potential drop across the resistor and I_{ds} is the drain current that is proportional to gate width. Using the scale factor s_1 defined above, parasitic source and drain resistances can be scaled from the following expressions:

$$R'_s = R_s/s_1 \tag{2.10}$$
$$R'_d = R_d/s_1 \tag{2.11}$$

The gate resistance in a MESFET or HEMT is proportional to the ratio Z/N^2 [1] where Z is the gate width and N is the number of gate fingers. This relation arises from the fact that each gate finger has resistance proportional to Z/N, with N parallel fingers. To scale gate resistance, a different scale factor must be defined:

$$s_2 = \frac{Z'/N'^2}{Z/N^2} \tag{2.12}$$

from which R_g is scaled to its new value as

$$R'_g = R_g s_2 \tag{2.13}$$

Figures 2.3(a) through 2.3(f) show actual data extracted from measured s-parameters using the techniques described in Chapter 4 for four devices of different sizes: 150 μm $\times$ 4 fingers, 300 μm $\times$ 4 fingers, 600 μm $\times$ 8 fingers, and 900 μm $\times$ 8 fingers. The data are in excellent agreement with the scaling rules, demonstrating the capability of determining small-signal models for many different size devices from measurements of a single device.

2.1.3 Physically Based Modeling

In this section, the physically based analytical model of Pucel is presented [2]. Because it is based on fundamental physical parameters, it is useful in analyzing the effects of gate length variation, doping variation, and epi-thickness variation on small-signal parameters such as transconductance, gate-source capacitance, *et cetera*. The model is also useful in analyzing devices made from materials other than GaAs because fundamental physical parameters such as mobility, saturation veloc-

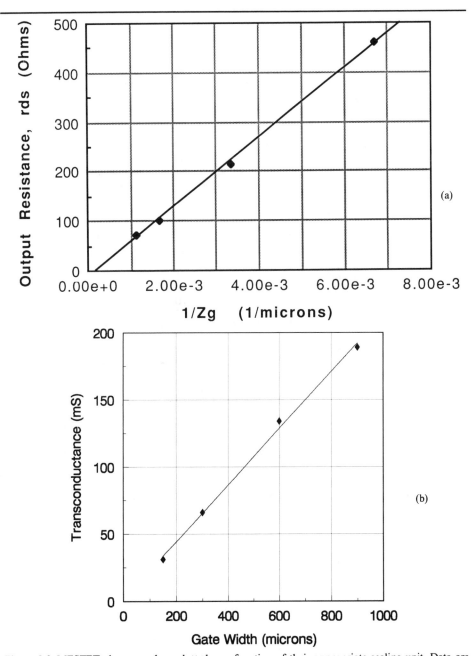

Figure 2.3 MESFET element values plotted as a function of their appropriate scaling unit. Data are extracted from measured *S*-parameters for four different size devices: 150 μm × 4 fingers, 300 μm × 4 fingers, 600 μm × 8 fingers, 900 μm × 8 fingers. (a) Output resistance; (b) transconductance; (c) gate-source capacitance; (d) gate-drain and drain-source capacitance; (e) parasitic drain and source resistance; (f) parasitic gate resistance.

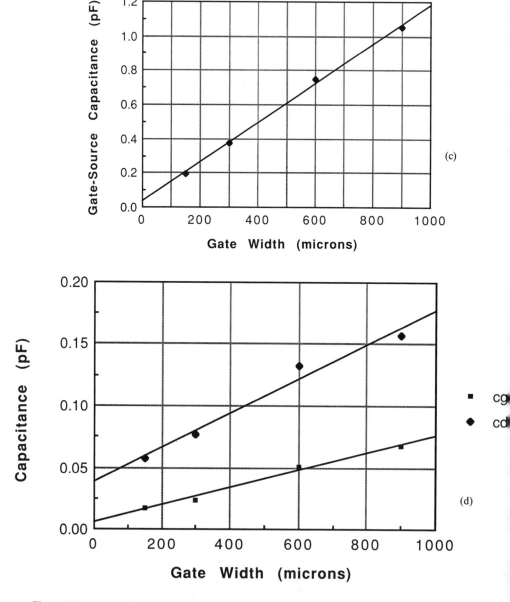

Figure 2.3 continued

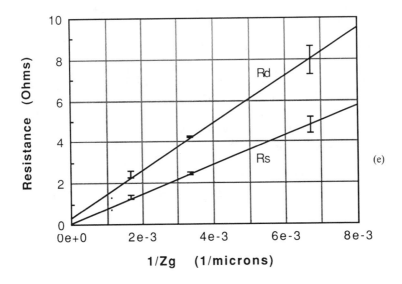

(e)

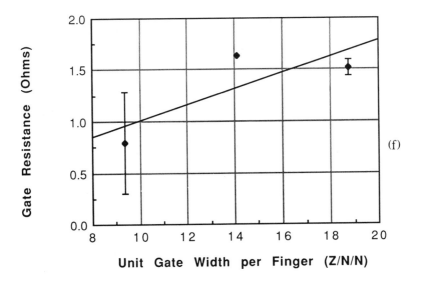

(f)

Figure 2.3 continued

ity, dielectric constant, *et cetera* are used in the model [1]. The model has also been extended for use in profile studies of ion-implanted MESFETs [3, 4].

2.1.3.1 Direct Current Analysis

This analysis is based on the idealized symmetric FET structure depicted in Figure 2.4. The channel is divided into two longitudinal sections, $x < L_1$ where the electric field is proportional to the mobility μ_0 and $L_1 \leq x \leq L$ where velocity saturation is assumed. Because the model makes an abrupt transition from ohmic behavior to a saturated condition, it is sometimes referred to as the *two-region model*. The velocity field curve for the model is compared in Figure 2.5 to the velocity-field curve predicted by Monte Carlo simulations.

For the purpose of this analysis, we define a channel potential $W(x)$ with respect to the gate electrode

$$W(x) = V_{gs} + V_{bi} - V(x), \tag{2.14}$$

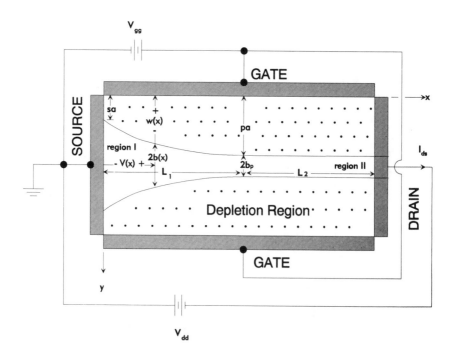

Figure 2.4 Cross-sectional diagram of FET showing various geometrical dimensions and potentials used in the Pucel analysis.

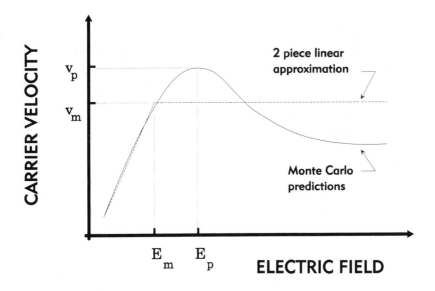

Figure 2.5 GaAs velocity-field curve for the two-region Pucel approximation.

where $V(x)$ is the channel potential with respect to the source, V_{gs} is the gate-source bias potential, and V_{bi} is the gate junction built-in potential. Evaluating this channel potential at the source, drain, and pinch-off points gives the following expressions:

$$W_s = V_{gs} + V_{bi} \tag{2.15}$$

$$W_d = V_{gs} + V_{bi} - V_{ds} \tag{2.16}$$

$$W_p = V_{gs} + V_{bi} - V_p \tag{2.17}$$

Further notation is simplified by definition of the following normalized potentials:

$$s = (W_s/W_{00})^{1/2} \tag{2.18}$$

$$p = (W_p/W_{00})^{1/2} \tag{2.19}$$

$$d = (W_d/W_{00})^{1/2} \tag{2.20}$$

$$w(x) = [W(x)/W_{00}]^{1/2} \tag{2.21}$$

where

$$W_{00} = (qN_d a^2/2\epsilon_r\epsilon_0) \tag{2.22}$$

is the gate-to-channel potential required to deplete the channel of carriers, a is the epitaxial layer thickness of the planer device (half of the device in Figure 2.4), N_d is the doping density in the channel, $\epsilon_r = 12.5$ is the relative dielectric constant of GaAs, $\epsilon_0 = 8.854 \times 10^{-14}$ F/cm is the permittivity of free space, and $q = 1.602 \times 10^{-19}$ C is the electron charge.

The channel-gate potential is obtained by the one-dimensional solution of Poisson's equation in the transverse direction. This approximation is justified on the basis that the longitudinal field in the channel is negligible compared to the transverse field in the depletion region (this approximation breaks down for some improperly scaled short-channel devices). The boundary condition used for the solution is that the electric field vanishes at the depletion region edge (abrupt depletion approximation). Under these conditions, the channel potential $W(x)$ is given by

$$W(x) = W_{00}[1 - b(x)/a]^2 \tag{2.23}$$

where $2b(x)$ is the dimension of the undepleted opening in the channel at x.

The drain current for the device is

$$I_{ds} = 2b(x)Z\sigma E_x(x) \tag{2.24}$$

where $E_x(x) = dW/dx$ is the longitudinal channel field, Z is the gate width, $\sigma = q\mu_0 N_d$ is the epi layer conductivity and μ_0 is the low field mobility. Integrating (2.24) from $x = 0$ to $x = L_1$ using the normalized potentials defined in equations (2.18) through (2.21) yields the following expression for the drain current of the symmetric FET [2]:

$$I_{ds} = \frac{g_0 Z W_{00}}{L_1} f_1(s,p) \tag{2.25}$$

where

$$f_1(s,p) = p^2 - s^2 - \tfrac{2}{3}(p^3 - s^3) \tag{2.26}$$

and

$$g_0 = 2a\sigma \tag{2.27}$$

is twice the sheet conductivity of the epitaxial layer of the asymmetric FET. The factor of 2 in (2.27) is included to make expressions including the quantity g_0 valid for the symmetric FET. By removing the 2 in (2.27), the expressions cited in this subsection are equally valid for asymmetric FETs.

The quantity L_1 can be determined by assuming current continuity between region I (ohmic) and region II (saturated). The current in the velocity saturated region is given by

$$I_{ds}(\text{region II}) = I_{dsat} = g_0 Z E_s(1 - p) \qquad (2.28)$$

Equating equations (2.25) and (2.28) at the discontinuity in the two-piece linear approximation for the velocity field curve gives for L_1:

$$L_1 = L[f_1(s,p)/\xi(1 - p)] \qquad (2.29)$$

where $\xi = E_s L/W_{00}$.

The source-to-drain potential drop is obtained by integrating the longitudinal electric field along the channel from $x = 0$ to $x = L_1$ and from $x = L_1$ to $x = L$. The resulting expression for V_{ds} is given by [2]:

$$V_{ds} = -W_{00}\left[(p^2 - s^2) + \frac{2}{\pi}\left(\frac{a}{L}\right)\xi \sinh\frac{\pi L_2}{2a}\right]. \qquad (2.30)$$

where $L_2 = L - L_1$.

Equation (2.30) may be combined with (2.29) to eliminate L_1 and solve for p, given the applied potentials V_{gs} and V_{ds}. The drain current is then easily obtained in either region from equation (2.25) or (2.28).

2.1.3.2 Transconductance and Output Conductance

The transconductance is obtained by differentiating equation (2.28) with respect to V_{gs}. Doing so gives for transconductance,

$$g_m = -dI_{ds}/dV_{gs} = (I_s/W_{00})f_g(s,p,\xi) \qquad (2.31)$$

where f_g is a nondimensional function given by

$$f_g(s,p,\xi) = \frac{(1 - s)\cosh(\pi L_2/2a) - (1 - p)}{[2p(1 - p) + \xi(L_1/L)]\cosh(\pi L_2/2a) - 2p(1 - p)} \qquad (2.32)$$

and $I_s/W_{00} = 4\epsilon_r\epsilon_0 v_{sat}Z/a$.

For short-channel devices, when the approximations $s = p$ and $L_1 = 0$ are valid, the expression for g_m simplifies to

$$g_m \approx I_s(2pW_{00}). \qquad (2.33)$$

The output resistance is given by

$$r_{ds} = -dV_{ds}/dI_{ds} = (W_{00}/I_s)f_r(s,p,\xi)$$
(2.34)

where

$$f_r(s, p, \xi) = \frac{1}{1 - p}\left\{\left[2p(1 - p) + \xi\frac{L_1}{L}\right]\cosh\pi\frac{L_2}{2a} - 2p(1 - p)\right\}$$
(2.35)

For short gate lengths and moderate drain currents, the expression for r_{ds} simplifies to

$$R_{ds} = \frac{\pi L|V_{ds}|}{\xi a I_s}p$$
(2.36)

where $I_s = g_0 Z E_s$.

2.1.3.3 Capacitances C_{gs}, C_{gd}, and C_{ds}

The gate-source capacitance for the symmetric FET is given by [2]:

$$C_{gs} = 2\epsilon_r\epsilon_0 Z f_c(s,p,\xi)$$
(2.37)

where $f_c = f_{c1} + f_{c2} + 1.56$ and

$$f_{c1}(s,p,\xi) = \frac{2}{f_1}\frac{L_1}{a}\left\{f_g\left[\frac{2p^2(1 - p)^2 + f_2}{1 - p}\right] - s(1 - s)\right\}$$
(2.38)

$$f_{c2} = (s,p,\xi) = 2\frac{L_2}{a}f_g + (1 - 2pf_g)\left[2\frac{L}{a}\frac{p}{\xi\cosh(\pi L_2/2a)} + \tanh\frac{\pi L_2}{2a}\right]$$
(2.39)

The expressions f_{c1} and f_{c2} represent the contributions from regions I and II while the numerical term accounts for fringing capacitance at either edge of the gate electrode. The expression for gate-source capacitance simplifies considerably for short gate lengths and gate biases that are not too near the pinch-off voltage. Under this approximation, $s = p$, $L_1 = 0$, and we obtain

$$C_{gs} = 2\epsilon_r\epsilon_0 Z(L/ap + 1.56).$$
(2.40)

To first order, both C_{gd} and C_{ds} are independent of bias. For the asymmetric FET, they can both be expressed by the form

$$C_{gd}, C_{ds} = (\epsilon_r + 1)\epsilon_0 Z \frac{K(1 - k^2)^{1/2}}{K(k)} \qquad (2.41)$$

where $K(k)$ is the complete elliptic integral of the first kind. The argument k for the respective capacitances is given by

$$k_{gd} = \left[\frac{L_{gd}}{L_{gd} + L} \right]^{1/2} \qquad (2.42)$$

$$k_{ds} = \left[\frac{(2L_s + L_{ds})L_{ds}}{(L_s + L_{ds})^2} \right]^{1/2} \qquad (2.43)$$

where L_{gd} and L_{ds} are the interelectrode spacings depicted in Figure 2.6.

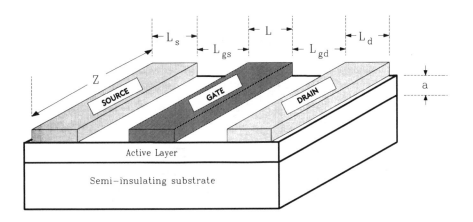

Figure 2.6 Perspective of FET structure showing important dimensions.

2.1.3.4 Pucel Model Results

As discussed in the opening of this section, a physically based model such as the Pucel model can be used to analyze the effects of the variation of process-dependent parameters on small-signal parameters. Table 2.1 gives the physical parameters of a hypothetical GaAs MESFET. These parameters have been inserted into the Pucel model to obtain predictions of drain current, transconductance, and output con-

Table 2.1
MESFET Model Parameter Values Used for the Analysis Illustrated in Figures 2.7 through 2.9

a	0.173 μm
L	0.653 μm
Z	300 μm
N_d	1.29×10^{17} cm^{-3}
μ_0	0.45 m/V-s
ν_{sat}	1.3×10^7 cm/s

ductance for varying dopants, gate lengths, and channel thicknesses. The nominal circuit parameters obtained with these physical inputs are I_{ds} = 29.5 mA, g_m = 31.2 mS, and g_{ds} = 0.325 mS. The variations of the circuit parameters are plotted as a function of doping, channel thickness, and gate length in Figures 2.7 through 2.9, respectively. The circuit parameters are normalized to the nominal values given in the previous paragraph while the physical parameters are normalized to the values given in Table 2.1.

2.1.4 Modeling Frequency-Dependent Conductances

Small-signal models such as the one shown in Figure 2.1 are usually adequate for microwave design because the output conductance dispersion occurs at frequencies

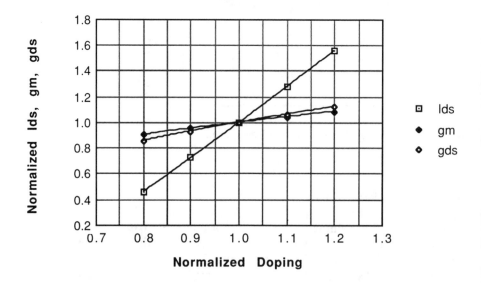

Figure 2.7 Variation of I_{ds}, g_m, and g_{ds} as a function of channel doping.

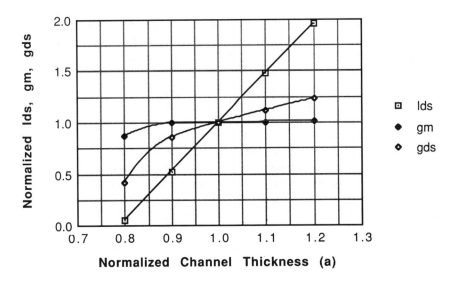

Figure 2.8 Variation of I_{ds}, g_m, and g_{ds} as a function of channel thickness.

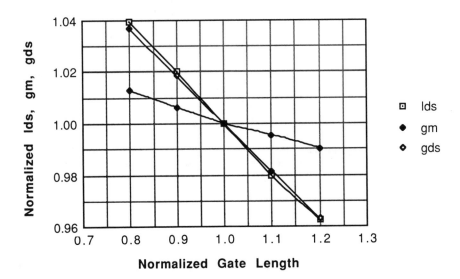

Figure 2.9 Variation of I_{ds}, g_m, and g_{ds} as a function of gate length.

well below the design frequencies. For large-signal modeling or low frequency applications, however, both the dc (low frequency) response and the high frequency response must be described. In these cases, the equivalent circuit in Figure 2.1 is inadequate.

The dispersive nature of MESFET operation can be modeled with the linear equivalent circuit in Figure 2.10 [5]. The additional current source shown in the figure models current injected into trapping states. The trapping states are located primarily at the surface, but also at the channel-substrate interface of the device. This current is coupled into the channel through the capacitance C_{ss}. For dc signals, the impedance across the capacitance C_{ss} is infinite and device transconductance and output conductance are given by the partial derivatives of the dc drain current expressed in equations (2.1) and (2.2). Under microwave operation, however, the transconductance and output conductance are functions of the dc values as well as the additional elements g_{m2} and R_{ss}. With proper selection of element values, this equivalent circuit is capable of modeling both low and high frequency conductances as well as the transition from the low to high frequency response.

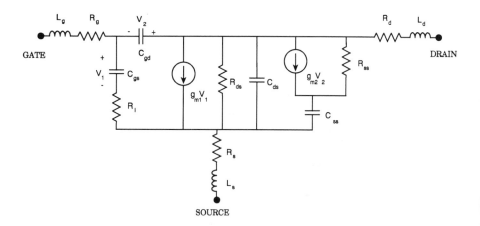

Figure 2.10 Small-signal equivalent circuit model of the FET including dispersion elements g_{m2}, R_{ss}, and C_{ss}. Note that output resistance, R_{ds}, represents the dc resistance for this model.

2.2 NOISE MODELS

Noise in electrical systems is the internal generation of signals that cause degradation from the desired or theoretical response. Fluctuations in signal phase, amplitude, and spectral content are forms of noise. The physical properties of materials result in various classes or types of noise, including white, phase, *et cetera.* For the MESFET and HEMT, particularly in amplifiers, discrete devices, and mixers, white noise is a critical concern. The term *white* refers to the distribution in the spectral content.

Thermal noise originates because heat in the electrical device provides energy to the carriers causing random fluctuations in their movement. This noise is generated only in systems or circuit elements that dissipate power (resistive); purely reactive elements (ideal capacitors, inductors, and transmission lines) do not generate noise. The rms thermal voltage produced by a thermal source such as a resistor R is given by [6–8]:

$$V_N = (4kTBR)^{1/2} \tag{2.44}$$

where

k = Boltzmann's constant (1.374×10^{-23} J/K),
T = temperature in kelvins, and
B = bandwidth in hertz.

This relationship is important in the development of noise figure measurements and the predictions of noise properties of two-port networks. The noise voltage described in equation (2.44) can also be discussed in terms of power. The maximum power generated by a resistor (Figure 2.11) occurs for a matched load and is given by the simplified expression

$$P_N = \frac{V_N^2}{4R} = kTB \tag{2.45}$$

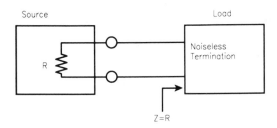

Figure 2.11 The maximum transfer of noise power generated in a resistor requires a matched load.

Note that the expression on the far right does not include the resistance value R. Because it is derived assuming the resistor is matched to the load, the factor R is eliminated from the equation. When the load is not matched to the resistor (Figure 2.12), the power delivered to the load is less than maximum and is given by

$$P_N = (1 - |\Gamma|^2)kTB \tag{2.46}$$

where Γ is the voltage reflection coefficient.

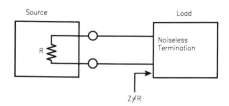

Figure 2.12 An unmatched load reflects noise power.

The noise characteristics of two-port networks can be derived by considering the network of Figure 2.13. A signal S_i with some noise N_i is present at the input of the device. Both the signal and noise will be affected by the system such that at the output a signal S_o and noise N_o are present. A figure of merit that quantitatively describes this behavior is the noise factor F of the system. The quantity F is defined as

$$F = \frac{S_i/N_i}{S_o/N_o} = \frac{N_o}{GN_i} \tag{2.47}$$

where

$\quad S_{i,o}$ = signal power at the input and output ports,
$\quad N_{i,o}$ = noise power at the input and output ports, and
$\quad G$ = two-port power gain.

Figure 2.13 A two-port network with both signal and noise present at input and output.

Note that under ideal conditions the two-port device will add no noise of its own to the signal. Rather, it simply amplifies or attenuates the signal and the input noise. Under these conditions, the noise factor F takes the value of 1 ($F = 1$). More commonly, however, this noise property is expressed as *noise figure,* which is given as

$$F_{dB} = 10 \log (F) \tag{2.48}$$

Another important parameter used to define noise characteristics is the noise temperature T_N. The relationship between noise temperature and noise figure is given by

$$F_{dB} = 10 \log (1 + T_N/T_0) \tag{2.49}$$

where T_0 is room temperature (290 K).

The noise temperature of a system should not be confused with a temperature of the system that is measurable with a thermometer. Instead, the noise temperature is the equivalent thermal energy required for an ideal resistor, matched to the input of a perfectly noiseless two-port device, to produce the same output noise.

In nonideal systems (systems containing resistors that act only as thermal sources), the relationship between input and output noise and signals is easily expressed in terms of noise factor F [6]:

$$F = 1 + \frac{T_N}{T_0} \left(\frac{1}{G} - 1 \right) \tag{2.50}$$

In systems containing only thermal sources, the input and output noise is the same ($N_i = N_o$), while the output signal S_o is attenuated.

When a two-port component contains active devices such as a MESFET, HEMT, or diode, noise is generated by these devices and is added to the output of the two-port device. The generation of this noise is typically modeled using an equivalent circuit for the device that contains current or voltage noise sources to model the noise effects. These internal noise sources contribute only to the overall noise of the two ports and not to the gain. However, because these sources are internal to the device, the overall noise is affected by the matching circuitry connected to the MESFET or HEMT.

One of the key factors influencing the noise figure of an amplifier circuit or active device is the generator admittance, $Y_g = G_g + jB_g$, connected at the input port of the device under test (DUT). This admittance is illustrated in Figure 2.14. The effect this has on noise figure is given by [6]:

$$F(Y_g) = F_{min} + \frac{R_n}{G_g}[(G_g - G_{opt})^2 + (B_g - B_{opt})^2]$$ (2.51)

where

F_{min} = minimum value of F with respect to Y_g,
$Y_{opt} = G_{opt} + jB_{opt}$ is the admittance value at which $F = F_{min}$, and
R_n = equivalent noise resistance of the DUT.

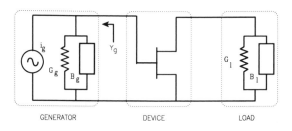

GENERATOR   DEVICE   LOAD

Figure 2.14 The circuitry connected at the input of a device significantly influences the device noise performance.

Equation (2.51) can also be expressed in terms of the generator or source reflection coefficient. The reflection coefficient is related to the admittance by:

$$\Gamma_g = \frac{Y_R - Y_g}{Y_R + Y_g}$$ (2.52)

where Y_R is the reference admittance for Γ.

A similar expression can be written for the optimal source reflection coefficient, Γ_{opt}, which results in a minimum noise figure. Hence, equation (2.51) can be written as

$$F = F_{min} + \frac{4R_n|\Gamma_g - \Gamma_{opt}|^2}{R_R(1 - |\Gamma_g|^2)|1 + \Gamma_{opt}|^2}$$ (2.53)

Equations (2.51) through (2.53) provide the basis for evaluating the noise figure when external circuitry is connected to the input of a device.

The meaning of the parameters Γ_{opt} and F_{min} are obvious from their stated definitions. The parameter R_n is useful as an indication of the dependence of F_{min}

on the source admittance Y_g. A low value of R_n is desirable for most circuit applications because this allows a source admittance considerably different from Y_g to be used without seriously degrading the noise figure.

The noise parameters F_{min}, F_n, and Γ_{opt} are properties only of the active device and are independent of any external circuitry connected to the device. They are, however, dependent on other parameters. For MESFETs and HEMTs, the other important parameters include temperature, frequency, drain-source current, and, to a much lesser extent, drain-source voltage. They are also dependent on the physical properties of the device, including geometric as well as material properties of the structure.

The remainder of this section discusses several methods of modeling noise properties of MESFETs and HEMTs. Several models are discussed—both physical and empirical. Discussions focus on several models available to determine parameters F_{min}, R_n, and Γ_{opt} for a MESFET or HEMT, and the dependence of these parameters on geometric dimensions, material characteristics, or bias of the device. A number of models, including empirical, semiempirical, and physical, have been proposed; all require an equivalent circuit for the device. To this end, the MESFET and HEMT equivalent circuit models and techniques to extract the model parameters described in Chapters 2, 3, and 4 are required.

2.2.1 Empirical Noise Models

An analysis that limits the number of measurements needed to obtain noise figure predictions was demonstrated by Podell in 1981 [9]. In this case, one noise measurement and a small-signal equivalent circuit model for the FET are sufficient for the complete noise analysis. More recently, Gupta *et al.* [10] described a model that further simplifies the measurement requirements while improving the noise figure predictions. This model also eliminates assumptions required of the earlier work related to circuit losses, noise correlations, and the functional form of the output noise current. This model, as well as most other empirical models, also describes HEMT noise behavior.

Figure 2.15 illustrates the equivalent circuit model for an amplifier. It consists of a generator of admittance, $Y_g = G_g + jB_g$; an input matching circuit of admittance, $Y_c = G_c + jB_c$, and the FET. In developing the noise model, the MESFET or HEMT device is represented by a five-element equivalent circuit model using gate-source capacitance C_{gs}; input resistance $R_T = R_g + R_s + R_i$; transconductance g_m; output resistance r_{ds}; and a white noise current source of spectral density S_{io}. The first four required equivalent circuit element values may be determined using S-parameter measurements and the small-signal analysis described in Chapters 2, 3, and 4. Evaluation of the spectral density of the white noise current source requires one noise measurement. This measurement is made directly using a low

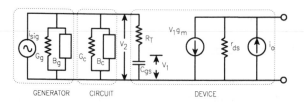

GENERATOR   CIRCUIT   DEVICE

Figure 2.15 An equivalent circuit of a FET amplifier consisting of a generator, input circuit, and device.

noise receiver, a bandpass filter and spectrum analyzer, or a bandpass filter and power meter. Details of these measurements are given in Section 3.4. The noise power is measured at the output of the device with the gate-source terminals of the device short circuited. The noise power generated by the device is dissipated in both the output resistance (r_{ds}) and the input resistance of the receiver (R_r). From Figure 2.16, the noise power spectral density generated in the device can be calculated given the measured noise power at the receiver P_{out} and the effective noise bandwidth B. This value is determined from the relationship [10]:

$$S_{io}(f_L) = \frac{P_{out}(R_r + r_{ds})^2}{Br_{ds}^2 R_r} \text{ A}^2/\text{Hz} \tag{2.54}$$

where f_L is the measurement frequency and R_r is the input resistance of the measurement equipment (receiver, spectrum analyzer, or power meter).

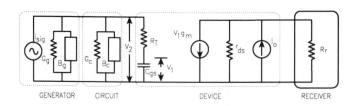

GENERATOR   CIRCUIT   DEVICE   RECEIVER

Figure 2.16 The noise of an amplifier circuit can be represented by an equivalent noise source between the drain-source terminals.

When the measurement frequency f_L is chosen high enough that low frequency effects such as $1/f$ noise are negligible, the power spectral density S_{io} is frequency independent—particularly in the microwave frequency range. This allows f_L to be chosen at a frequency convenient for the measurement equipment requirements. The spectral current density S_{io} is, however, dependent on the bias state of the device. Thus, the noise predictions from the Gupta model are limited to the device bias conditions in which S_{io} is measured.

A simplification in developing the model results by assuming the operating frequency f_o is significantly below the gain-bandwidth product of the device:

$$\omega_o^2 C_{gs}^2 R_T^2 \ll 1. \tag{2.55}$$

Also, for low noise, high quality devices, the gate leakage current is small and, when neglected, suggests that the uncorrelated noise conductance is equal to the input conductance of the MESFET or HEMT. Given these conditions and the equivalent circuit model shown in Figure 2.16, the input conductance and susceptance of the device is expressed as [10]:

$$G_{in} = \frac{\omega^2 C_{gs}^2 R_T}{1 + \omega^2 C_{gs}^2 R_T^2} \tag{2.56}$$

and

$$B_{in} = \frac{\omega C_{gs}}{1 + \omega^2 C_{gs}^2 R_T^2} \tag{2.57}$$

Using both the equivalent circuit element values and the power spectral density computed in equation (2.54), the noise resistance R_n can be evaluated from [10]:

$$R_n = S_{io}(f_L) \left(\frac{1 + \omega^2 C_{gs}^2 R_T^2}{4kT g_m^2} \right) \tag{2.58}$$

The optimal input circuit conductance and susceptance are given by the expression [10]:

$$G_{opt} = \left(G_{in}^2 + \frac{G_{in}}{R_n} \right)^{1/2} \tag{2.59}$$

and

$$B_{opt} = - B_{in} \tag{2.60}$$

The minimum noise figure is given as

$$F_{min} = 1 + 2R_n G_{in} + 2(R_n G_{in} + R_n^2 G_{in}^2)^{1/2} \tag{2.61}$$

The frequency-dependent properties F_{min}, R_n, and Y_{opt} for a given MESFET or HEMT are easily computed using equations (2.56) through (2.61). These values are constrained to the device bias condition in which S_{io} is measured. Based on these noise parameters, the noise figure for a circuit such as an amplifier containing a MESFET or HEMT is now computed using equation (2.51). This allows computation of the amplifier noise figure as a function of both frequency and source admittance.

Figures 2.17 through 2.19 present the equivalent circuit parameters C_{gs}, r_{ds}, and g_m for a 0.5×300-μm microwave MESFET as a function of drain current. These equivalent circuit parameters are obtained by extraction of S-parameter measurements at several drain-source current bias conditions. The input resistance ($R_T = R_G + R_i + R_S$) for this particular device is 5.0 Ω.

The output noise power of a MESFET or HEMT can be measured with the gate-source terminals RF short circuited using a large capacitive reactance. Using a receiver with a 50-Ω input impedance, the noise power at the drain-source terminals can be measured. Table 2.2 presents measured data for a microwave MESFET at several bias conditions and at a center frequency of 1.6 GHz. These noise power measurements are shown normalized to 1 Hz bandwidth. Based on these noise power measurements, the current spectral density S_{io} is computed from equation (2.54) and is shown graphically in Figure 2.20. From equations (2.54) through (2.61), F_{min} is also computed for each drain-source current and the results

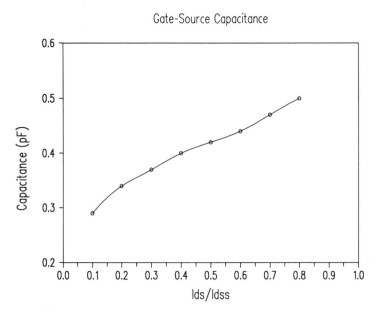

Figure 2.17 Extracted gate-source capacitance (C_{gs}) of a 0.5×300-μm MESFET as a function of drain current.

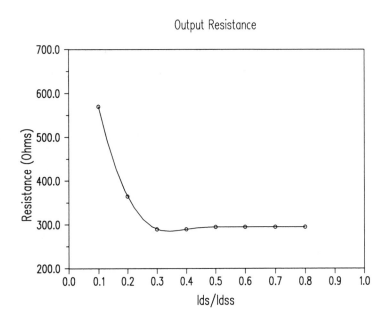

Figure 2.18 Extracted drain-source resistance (r_{ds}) of a 0.5×300-μm MESFET as a function of drain current.

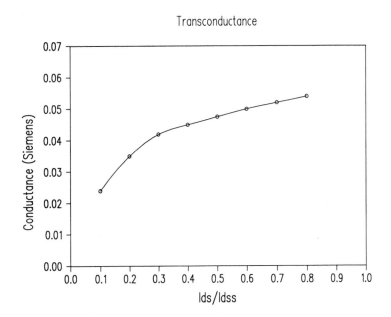

Figure 2.19 Extracted transconductance (g_m) of a 0.5×300-μm MESFET as a function of drain current.

Table 2.2
Noise Power and Current Spectral Density for a 0.5×300-μm MESFET

I_{ds}/I_{dss}	Device Noise Power W/Hz	S_{io} A^2/Hz
0.10	1.86×10^{-20}	4.41×10^{-22}
0.20	1.86×10^{-20}	4.81×10^{-22}
0.30	2.95×10^{-20}	8.11×10^{-22}
0.40	3.72×10^{-20}	1.02×10^{-21}
0.50	4.68×10^{-20}	1.28×10^{-21}
0.60	5.89×10^{-20}	1.61×10^{-21}
0.70	8.13×10^{-20}	2.22×10^{-21}

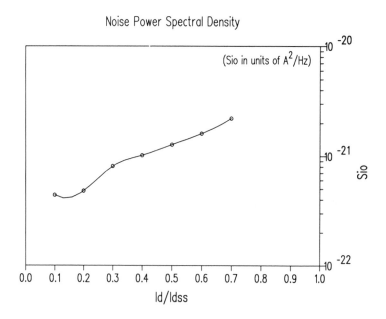

Figure 2.20 Noise power spectral density of a 0.5×300-μm MESFET measured at the drain-source terminal with the gate-source terminals short circuited.

are shown in Figure 2.21. Similarly, the noise parameters Γ_{opt}, R_n, and F_{min} were calculated from 2 to 18 GHz at a drain-source current of $0.5 I_{dss}$. These results are compared to those made with the direct on-wafer measurement technique described in Section 2.2.1 and are shown graphically in Figure 2.22.

An inspection of these results suggests good agreement. However, these results are sensitive to the value of R_T used in the model. This value includes both R_g and

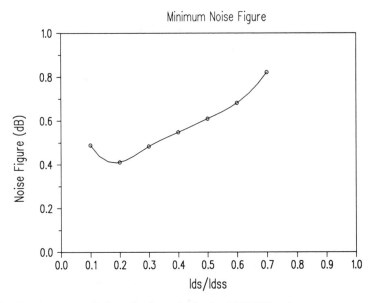

Figure 2.21 Based on an equivalent circuit model for the MESFET and a measurement of the noise power spectral density, the Gupta model predicts the device minimum noise figure F_{min}.

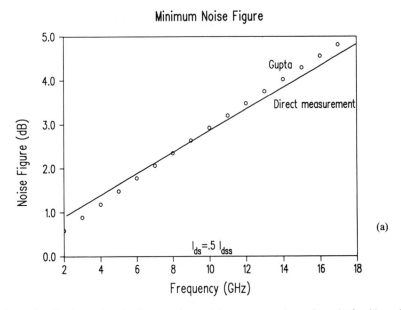

Figure 2.22 Predicted values using the Gupta noise model are compared to values obtained by a direct measurement: (a) F_{min}; (b) R_n; (c) R_{opt}; (d) X_{opt}.

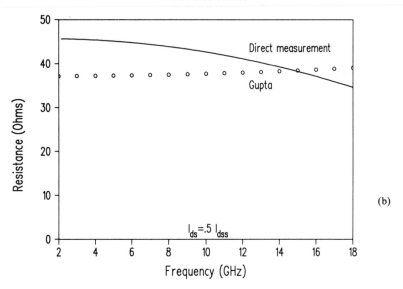

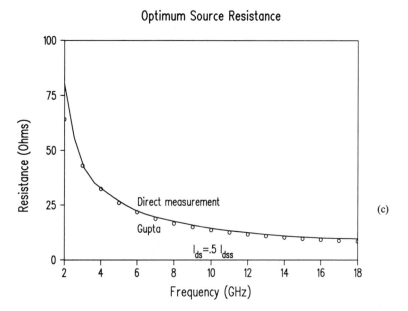

Figure 2.22 continued

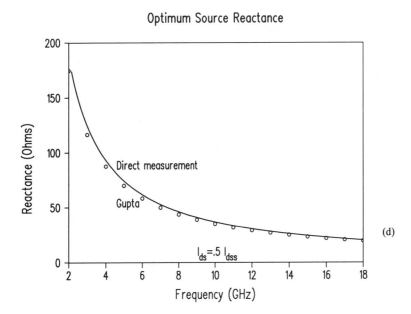

Figure 2.22 continued

R_i equivalent circuit element values and, as discussed in Section 2.1, these element values are sometimes difficult to determine accurately.

Another useful model for the prediction of noise characteristics of MESFETs and HEMTs was proposed by Fukui [11, 12]. By evaluating FET noise properties, Fukui develops several empirical equations that describe the frequency dependence of two-port MESFET noise parameters. Although these results are empirically derived, the model has proven to predict the noise characteristics of microwave MESFETs accurately and reliably. This model is of particular advantage for low-noise amplifier design applications, for which a convenient and quick method of predicting MESFET noise characteristics is needed.

The most basic form of the model describes the noise characteristics of a MESFET or HEMT as a two-port device (common source connection). The frequency dependence of parameters $F_{\min}$, Γ_{opt}, and R_n is predicted at a specified bias condition.

Similar to the Gupta method, one noise figure measurement and a small-signal equivalent circuit model of the device are required. In this case, the noise measurement must directly or indirectly define the $F_{\min}$, R_n, and Γ_{opt} of the MESFET or HEMT at one frequency. Also, the bias conditions of the device where this measurement is made define the bias conditions where the Fukui model applies. If a noise model is desired at a different bias condition, then these noise parameters must be measured at that bias state.

In developing these empirical relationships, the small-signal equivalent device model shown in Figure 2.23 is considered. This model is the same as the Pucel model (see Section 2.2.2), which models the intrinsic MESFET and parasitic resistances R_s and R_g and the charging resistance R_i. Any parasitic reactance on the drain, gate, or source terminals is not included. Although the noise model is empirical in nature, the small-signal element values can be related to physical parameters of the device as described by Pucel. Alternatively, and more precisely, these values can be determined using the characterization and extraction methods outlined in Chapters 2, 3, and 4.

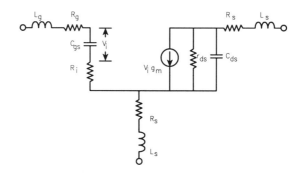

Figure 2.23 Equivalent model of a FET to predict noise performance based on the Fukui method.

The following empirically derived relationships are found to reliably predict the noise characteristics [11, 12]:

$$F_{min} = 1 + k_1 f C_{gs} \left(\frac{R_g + R_s}{g_m} \right)^{1/2} \tag{2.62}$$

$$R_n = k_2 / g_m \tag{2.63}$$

$$R_{opt} = \frac{k_3}{f} \left(\frac{1}{4g_m} + R_s + R_g \right) \tag{2.64}$$

$$X_{opt} = \frac{k_4}{f C_{gs}} \tag{2.65}$$

where k_1, k_2, k_3, and k_4 are empirical fitting factors and f is the operating frequency.

When C_{gs} is expressed in pF, resistance in ohms, conductance in siemens, and frequency in GHz, the value of k_1 is typically on the order of 0.01 to 0.02 for a microwave MESFET. Slightly lower values are obtained for high performance HEMTs.

The expression for R_{opt} of equation (2.64) has been modified from the Fukui work to account for the frequency dependence of this parameter. Characterization of several MESFETs using direct measurement techniques have shown the relationship of equation (2.64) to predict R_{opt} reliably. Measurements are also included later in this chapter that illustrate this frequency dependence.

Although it is not shown in the above equations, the value of R_g must incorporate the value of the charging resistance R_i if such an element is utilized in the equivalent circuit. Also, quite interestingly, equation (2.62) is identical to Pucel's physically derived minimum noise figure, equation (2.51), when the Pucel noise parameters $R = 0$ or $C = 1$.

To consider the accuracy of this model, the predicted noise parameters can be compared to those obtained by a direct noise parameter measurement technique. This has been accomplished using a microwave MESFET and some of the results are presented in Table 2.3 and Figure 2.24. The device was measured from 2 to 18 GHz using an on-wafer direct noise measurement technique. This measurement directly provides F_{min}, Γ_{opt}, and R_n at a bias condition of I_{dss} (i.e., with $V_{gs} = 0$ V and the device operating in current saturation conditions). In addition, from S-parameters measured at this same bias condition, small-signal element values for the noise model shown in Figure 2.23 are extracted. The extracted values are $g_m = 0.0684$ S; $R_g + R_i = 4.41\ \Omega$; $R_s = 1.73\ \Omega$; and $C_{gs} = 0.477$ pF. Noise parameters at drain-source bias currents of $0.2I_{dss}$, $0.5I_{dss}$, and I_{dss} were also measured and are shown in Figure 2.25. The data suggest that the Fukui model is applicable over a wide bias range.

Table 2.3
Evaluation of the Empirical Fitting Factors k_1, k_2, k_3, and k_4 where $I_{ds} = I_{dss}$

Frequency (GHz)	k_1	k_2	k_3	k_4
2.0	0.0335	3.08	14.97	147.2
4.0	0.0255	3.33	16.00	174.0
6.0	0.0236	3.39	15.33	179.3
8.0	0.0233	3.35	14.42	177.4
10.0	0.0236	3.20	13.94	172.1
12.0	0.0243	3.00	13.44	165.8
14.0	0.0253	2.73	13.74	158.3
16.0	0.0265	2.45	14.15	150.5
18.0	0.0279	2.16	14.61	139.1

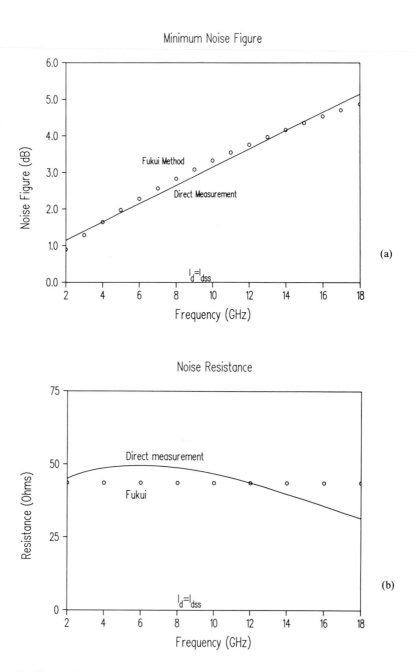

Figure 2.24 Predicted values using the Fukui noise model are compared to values obtained by direct measurement: (a) F_{min}; (b) R_n; (c) X_{opt}; (d) R_{opt}.

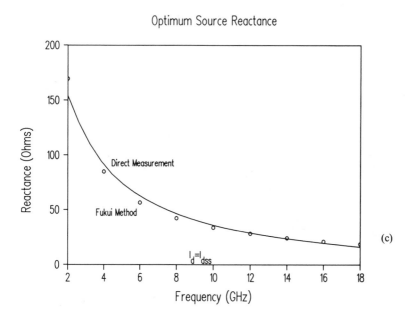

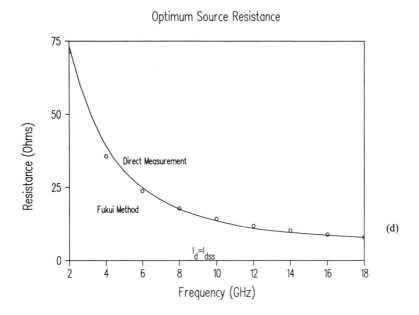

Figure 2.24 continued

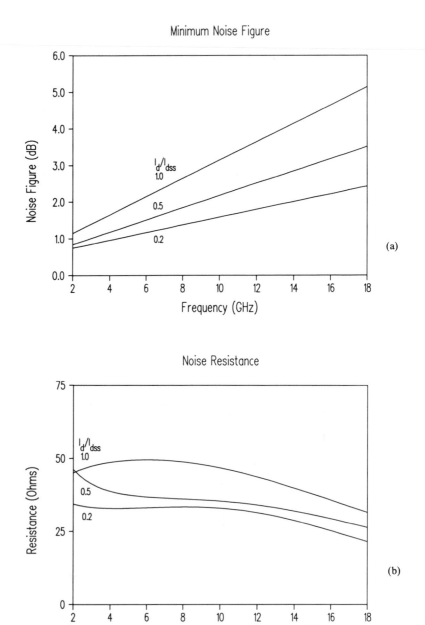

Figure 2.25 Measured noise parameters of a 0.5×300-μm MESFET at three values of drain source current: (a) F_{min}; (b) R_n; (c) X_{opt}; (d) R_{opt}.

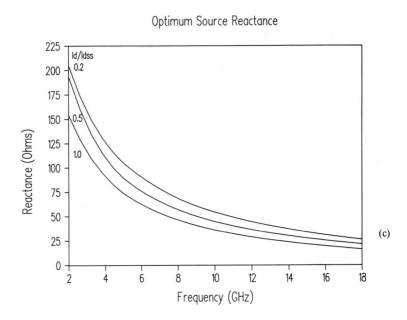

Optimum Source Reactance

(c)

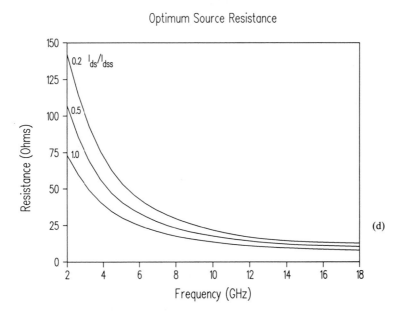

Optimum Source Resistance

(d)

Figure 2.25 continued

Equations (2.62) through (2.65) are written in terms of the empirical fitting factors, k_1 through k_4. Using measured noise parameters, these factors are computed from 2 to 18 GHz and are tabulated in Table 2.3. Although these factors should be independent of frequency, some variation is noted. Based on these results, an average value ($k_1 = 0.0259$, $k_2 = 2.966$, $k_3 = 14.51$, and $k_4 = 162.6$) is chosen for each fitting factor and the noise parameters are computed using equations (2.62) through (2.65). A comparison of these results to the direct measurements is shown graphically in Figure 2.24.

In addition to the MESFET characterization, noise parameters were measured for a 0.3×200-μm HEMT. These measurements are shown graphically in Figure 2.26 at three device bias conditions. The HEMT noise characteristics are seen to be very similar functionally to those presented for microwave MESFETs. Empirical noise models, therefore, can be applied to the HEMT as easily as they are applied to the MESFET.

Another important application of this model is that it allows a simple method of predicting the dependence of noise parameters by means of device geometry, such as gate width. When the noise parameters are known for a given FET of gate width Z, this model allows prediction of noise parameters for a FET scaled in width

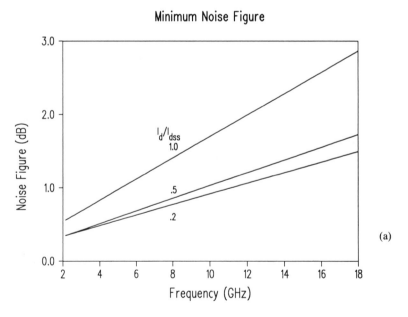

Figure 2.26 Measured noise parameters of a 0.3×200-μm HEMT at three values of drain source current: (a) $F_{\min}$; (b) R_n; (c) X_{opt}; (d) R_{opt}.

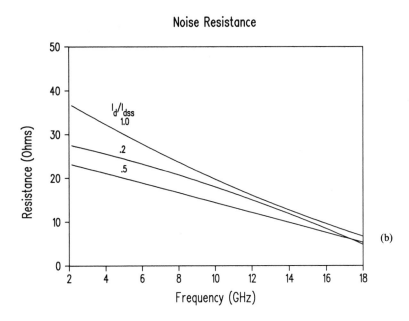

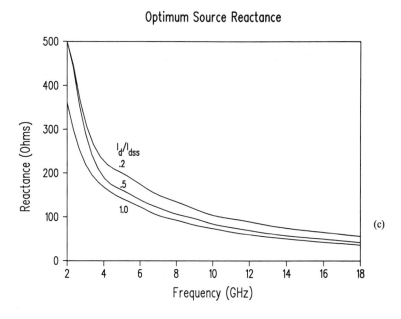

Figure 2.26 continued

Optimum Source Resistance

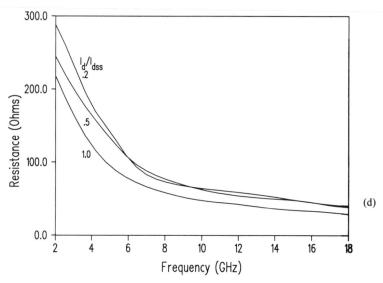

(d)

Figure 2.26 continued

to Z'. The empirical k factors are representative of material properties of the device as well as device bias conditions. To first order, simple expressions (developed in Section 2.1) relate R_g, R_s, g_m, and C_{gs} to gate width. Thus, for a constant drain bias (I_{ds}), these scaled circuit parameters are given by

$$R_{s'} = R_s/s_1 \tag{2.66}$$

$$g_{m'} = g_m s_1 \tag{2.67}$$

$$C_{gs'} = C_{gs} s_1 \tag{2.68}$$

and

$$R_{g'} = R_g s_2 \tag{2.69}$$

where s_1 and s_2 are scaling factors that relate a known FET of width Z to a scaled one of width Z' by the equations given in Section 2.1.

Thus, for a constant drain current, when noise parameters for a MESFET or HEMT of width Z are known, the scaled noise parameters for width Z' are obtained by substituting the above equations into equations (2.62) through (2.65). This leads to

$$R'_n = R_n/s_1 \tag{2.70}$$

$$X'_{opt} = X_{opt}/s_1 \tag{2.71}$$

$$F'_{min} = 1 + (F_{min} - 1) \left[\frac{1 + (R_g/R_s)(s_1 s_2)}{1 + (R_g/R_s)} \right]^{1/2} \tag{2.72}$$

$$R'_{opt} = R_{opt} \left[\frac{1 + (4g_m R_s) + (4g_m R_g\, s_1 s_2)}{1 + 4g_m R_s + 4g_m R_g} \right] \tag{2.73}$$

While equations (2.70) through (2.73) allow prediction of noise parameters as a function of gate width, prior knowledge of one such device is required.

To illustrate these concepts, the noise parameters for a 0.5×200-μm MESFET are predicted based on those of a 0.5×300-μm MESFET. Each FET is biased at $0.5 I_{dss}$ and fabricated using identical processes, with both devices fabricated on the same wafer. The noise parameters Γ_{opt}, F_{min}, and R_n for the 0.5×300-μm MESFET are measured by means of a direct measurement technique that allows computation of the k factors from equations (2.62) through (2.65). These noise parameters are measured using on-wafer probes from 2 to 18 GHz and k factors are computed at each frequency. The resulting factors are $k_1 = 0.0191$, $k_2 = 1.630$, $k_3 = 23.61$, and $k_4 = 164.3$. Based on these results, the scaling factors s_1 and s_2 are determined, and noise parameters for the 0.5×200-μm MESFET are calculated from equations (2.70) through (2.73). The predicted noise parameters R_{opt}, X_{opt}, R_n, and F_{min} for the scaled FET are shown graphically in Figure 2.27. For comparative purposes, the noise parameters of the 0.5×200-μm MESFET were also measured by means of a direct measurement technique, with the results shown graphically in Figure 2.27. Based on these measurements, the k factors are computed and found to be $k_1 = 0.0159$, $k_2 = 1.425$, $k_3 = 24.93$, and $k_4 = 176.3$ which are close in value to the k-factors for the $.5 \times 300$ μm MESFET.

The simplicity of this model and its ability to predict noise characteristics has been further investigated by Cappy *et al.* [13], who relate the empirical fitting factor k_1 in equation (2.62) to physical parameters of the device. Additionally, F_{min} is expressed in a form given by

$$F_{min} = 1 + k_f(f/f_T)[g_m(R_s + R_g)]^{1/2} \tag{2.74}$$

where k_f is an empirical fitting factor and f_T is the cutoff frequency given by

$$f_T = g_m/(2\pi C_{gs}). \tag{2.75}$$

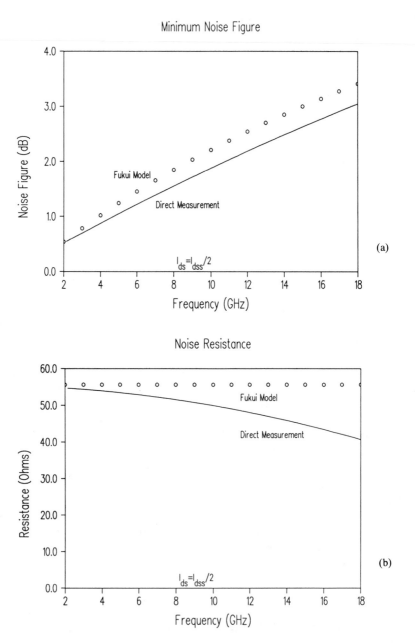

Figure 2.27 Noise parameters of a 0.5×200-μm FET are predicted based on those of a 0.5×300-μm FET using the Fukui noise model with results compared to those obtained by direct on-wafer measurements: (a) F_{min}; (b) R_n; (c) X_{opt}; (d) R_{opt}.

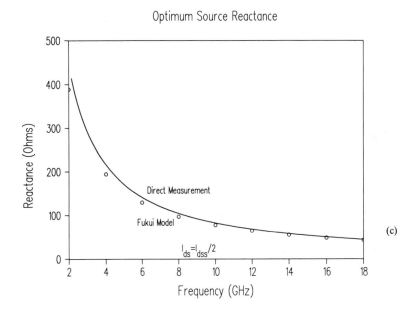

(c)

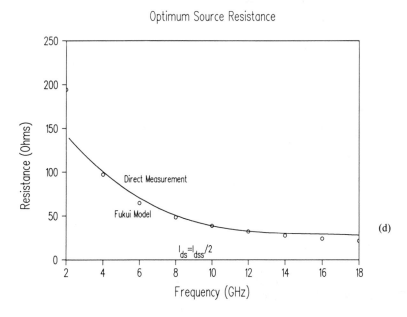

(d)

Figure 2.27 continued

Cappy relates the empirical fitting factor k_f to physical device characteristics of the MESFET, including gate width, gate length, and drain current I_{ds}. Considering equation (2.74) and Figure 2.23, the fitting factor can be expressed as

$$k_f = 2 \left(\frac{\langle i_d^2 \rangle}{4kTg_mB} \right)^{1/2} \tag{2.76}$$

where B is the bandwidth in hertz and i_d is the equivalent current generator for the drain noise given by

$$\langle i_d^2 \rangle = 4kTB\delta(\omega)(g_m/C_{gs})[L(\alpha Z + \beta I_{ds})] \tag{2.77}$$

and

$$\delta(\omega) = (g_{ds}^2 + \omega^2 C_{gd}^2)/g_{ds}^2 \tag{2.78}$$

where

$$
\begin{aligned}
Z &= \text{FET gate width,} \\
L &= \text{FET gate length,} \\
\alpha, \beta &= \text{fitting factors, and} \\
g_{ds} &= 1/r_{ds}.
\end{aligned}
$$

The expression for k_f further simplifies when $\delta(\omega)$ is near unity to

$$k_f = 2 \left[\frac{L}{C_{gs}} (\alpha Z + \beta I_{ds}) \right]^{1/2} \tag{2.79}$$

Substituting equation (2.79) into equation (2.74) results in an expression for the minimum noise factor:

$$F_{\min} = 1 + (8\pi)^{1/2} f \left[\frac{L}{f_T} (\alpha Z + \beta I_{ds})(R_s + R_g) \right]^{1/2} \tag{2.80}$$

2.2.2 Physical Models

The physical model proposed by Pucel *et al.* [2] is a comprehensive model of the MESFET that is derived from general principles in terms of fundamental physical parameters. Good agreement has been shown between measured and predicted results.

Pucel considers the general FET model shown in Figure 2.28. The model is simplified with appropriate current and voltage sources added to model noise effects (Figure 2.29). This model is derived by neglecting C_{gd}, C_{ds}, and the parasitic drain resistance R_d. These elements normally have a negligible effect on noise. The model assumes a common source configuration. Additionally, parasitic reactance on the drain, gate, and source terminals is neglected. Excluding the noise sources, the FET is modeled using gate-source capacitance C_{gs}, gate resistance R_g, charging resistance R_i, source resistance R_s, and transconductance g_m. These parameters may be derived from the expressions related to the physics of the MESFET given in Section 2.1. Alternatively, these parameters may be extracted from measurements as outlined in Chapter 4. The voltage source e_s represents an external generator of impedance Z_{sc}.

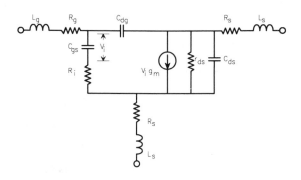

Figure 2.28 Equivalent circuit model for a FET.

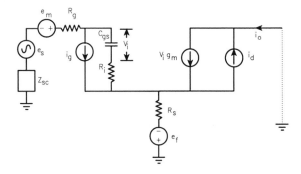

Figure 2.29 The Pucel noise model is based on a simplified FET model with noise sources represented by voltage and current sources.

This noise model considers both the intrinsic device as well as the thermal sources associated with the parasitic resistances (R_g, R_s, and R_d). The thermal source R_d is usually neglected because it follows the intrinsic FET and its contribution to overall device noise is negligible. Voltage sources e_m and e_f represent the thermal sources associated with parasitic resistances R_g and R_s. Current sources i_g and i_d model noise generated in the channel region.

Pucel accounts for the effect of each of these noise sources and then draws an equivalence to an intrinsic FET (noiseless) with appropriate noise generators at the input and output ports (Figure 2.29). This representation is particularly useful because it is closely related to the actual physical noise generation process of the FET. Noise current source i_d is identified with the channel noise generated under short-circuit drain conditions in the source-drain path, while the generator i_g can be related to the noise current induced in the gate circuit by charge fluctuations in the drain current. The two noise current sources, i_d and i_g, are also correlated, and this correlation is quantified.

The following discussion will consider the effects of each source—i_g, i_d, e_m, and e_f—and their physical relationship to the MESFET. A correlation coefficient C is defined that relates the current sources i_g and i_d. Finally, the parasitic resistance effects (R_g and R_s) and generator resistance R_{sc} are considered. An expression is then presented that relates the noise on the drain to the noise generators. Finally, an expression for noise figure F and the dependence on the source admittance are shown. Many of the physical parameters used in these expressions are defined in Section 2.1.

As described in Section 2.1, Pucel partitions the channel into two regions, a low field and a high field region. The drain circuit noise is caused by two phenomena: (1) thermal fluctuations and (2) carriers traveling at their saturated velocity, which is interpreted as diffusion noise. These two effects are uncorrelated because they are produced by different mechanisms. The time-average mean square voltage produced by these sources has been shown by Pucel et al. and Statz et al. [2, 14] to be given by

$$\overline{|V_{d1}^2|} = \frac{4kT_0B}{2a\sigma Z/L_1} \frac{P_0 + P_\delta}{(1-p)^2} \cosh^2 \frac{\pi L_2}{2a} \tag{2.81}$$

where

$$P_0 = (f)^{-1}[(p^2 - s^2) - (\tfrac{4}{3})(p^3 - s^3) + \tfrac{1}{2}(p^4 - s^4)] \tag{2.82}$$

and

$$P_\delta = 2\delta(f_1)^{-1}(1-p)^3 \left[(s-p) + \ln \frac{1-s}{1-p} \right] \tag{2.83}$$

The parameter δ is a noise temperature-electric field curve fitting factor, which has been determined empirically to take a value of about 0.64 for GaAs [1]. Other parameters of equations (2.81) through (2.83) are as defined in Section 2.1.

The noise voltage is proportional to bandwidth B. The kTB term of equation (2.46) further indicates the thermal properties of this noise. An empirical curve fitting factor δ in the P_δ term indicates that the effective noise temperature of the carriers is higher than that of the crystal in the presence of an applied field. The term P_o represents the Johnson fluctuations in the drain current.

Pucel [2] develops an expression for the diffusion noise given by

$$\overline{|V_{d2}^2|} = I_{ds} \left\{ \frac{64a}{\pi^5 V_m^3} \frac{qD_{hi}B}{(\epsilon_R\epsilon_0)^2 z^2} \frac{\sin^2[\pi(1-p)/2]}{(1-p)^2} \right\}$$
$$\times \left[e^{(\pi L_2/a)} - 4e^{(\pi L_2/2a)} + 3 + \frac{\pi L_2}{a} \right]$$

(2.84)

where I_{ds} is the drain current. The two voltages V_{d1} and V_{d2} are both impressed on the output resistance r_{ds}. The resultant current i_d is given by the expression

$$\overline{|i_d^2|} = \overline{|i_{d1}^2|} + \overline{|i_{d2}^2|}$$

(2.85)

where

$$i_{d1} = V_{d1}/r_{ds}$$

(2.86)

and

$$i_{d2} = V_{d2}/r_{ds}$$

(2.87)

The gate circuit noise is caused by fluctuations in the active channel. These fluctuations are capacitively coupled to the gate electrode and cause the channel current to be modulated, which then appears as noise at the output of the device. Again, considering the channel to be partitioned into low and high field regions, an expression for the short-circuit gate fluctuations produced in the low field mobility region of the channel is given by [2]:

$$|i_{g1}^2| = \omega^2 \left[\frac{16kT_0B}{a\sigma/(L_1Z)} \frac{(\epsilon_R\epsilon_0L_1)^2}{(\gamma a)^2} \right] (R_0 + R_\delta)$$

(2.88)

where

$$\gamma = \frac{(1-p)^2 f_r}{f_1 \cosh(\pi/2a)L_2}$$

(2.89)

$$R_0 = (f_1)^{-3}\{[(k'_n)^2(p^2 - s^2) - (\%)k'_n(k'_n + \gamma)(p^3 - s^3)]$$
$$+ \tfrac{1}{2}[(k'_n)^2 + 4k'_n\gamma + \gamma^2](p^4 - s^4) - (\%)[(k'_n\gamma - \gamma^2)(p^5 - s^5)$$
$$+ (\gamma^2/3)(p^6 - s^6)]\} \tag{2.90}$$

and

$$R_\delta = \delta(1 - p)^3(f_1)^{-3}[-2(k'_n - \gamma)^2\left(p - s + \ln\frac{1 - p}{1 - s}\right)$$
$$+ (2k'_n\gamma - \gamma^2)(p^2 - s^2) - (\%)\gamma^2(p^3 - s^3)] \tag{2.91}$$

The k terms are defined by the following expressions:

$$k_n = (f_1)^{-1}[(-\%)(p^3 - s^3) + (\%)(p^4 - s^4)$$
$$+ (s^2 - \%s^3)(p - s)] + \gamma \tag{2.92}$$
$$k'_n = k_n + (L_2/L_1)(1 - p) \tag{2.93}$$

The short-circuit gate current fluctuations produced by sources in the high field region are given by

$$|i_{g2}^2| = \omega^2 I_{ds}\left(\frac{64D_{hi}B}{\pi^5 v_m^5}\right)\left[\frac{L_1 k'_n(\gamma = 0)}{\epsilon_R \epsilon_0 a Z(1 - p)r_{ds}}\right]^2$$
$$\times \sin^2\left[\frac{\pi(1 - p)}{2}\right]\left[\exp\left(\frac{\pi L_2}{a}\right) - 4\exp\left(\frac{\pi L_2}{2a}\right)\left(+ 3 + \frac{\pi L_2}{a}\right)\right] \tag{2.94}$$

The interdependence between the drain and gate voltage fluctuations is described in terms of a correlation coefficient c, which is defined by

$$jC = \overline{i_g^* i_d}/(|\overline{i_g^2}| \, |\overline{i_d^2}|)^{1/2} \tag{2.95}$$

Pucel *et al.* [2] have expressed this coefficient as

$$C = \frac{S_0 + S_\delta}{[(R_0 + R_\delta)(P_0 + P_\delta)]^{1/2}}\left(\frac{P_1 R_1}{PR}\right)^{1/2} + \left(\frac{P_2 R_2}{PR}\right)^{1/2} \tag{2.96}$$

where

$$S_0 = (f_1)^{-2}\{k'_n[(p^2 - s^2) - (\%)(p^3 - s^3) + \tfrac{1}{2}(p^4 - s^4)]$$
$$+ \gamma[(\%)(p^3 - s^3) + (p^4 - s^4) - (\%)(p^5 - s^5)]\}, \tag{2.97}$$

and

$$S_\delta = 2\delta(f_1)^{-2}(1-p)^3 \left[(k'_n - \gamma)\left(s - p + \ln\frac{1-s}{1-p}\right) + \tfrac{1}{2}\gamma(p^2 - s^2) \right] \quad (2.98)$$

A set of dimensionless noise coefficients, P and R, is now introduced to characterize the drain and gate circuit noise, respectively;

$$P = P_1 + P_2 \quad (2.99)$$

$$P_1 = \frac{1-p}{f_1 f_g \gamma^2}(P_0 + P_\delta) \quad (2.100)$$

$$P_2 = \frac{1-p}{\xi f_r^2 f_g}(L/a)f_3 \quad (2.101)$$

Also,

$$R = R_1 + R_2 \quad (2.102)$$

where

$$R_1 = 4(L/a)^2 \left(\frac{f_1}{1-p}\right)^3 \left(\frac{1}{\gamma\xi}\right)^2 \frac{f_g}{f_c^2}(R_0 + R_\delta) \quad (2.103)$$

and

$$R_2 = 4(L/a)^3 \left[\frac{1}{\xi(1-p)}\right]^3 \frac{f_1^2 f_g}{f_r^2 f_c^2}[k'_n(\gamma = 0)]^2 f_3 \quad (2.104)$$

Finally, f_3 is defined as

$$f_3 = \frac{16D_{hi}}{\pi^3 D_0} \left[\frac{\sin(\pi/2)(1-p)}{(\pi/2)(1-p)}\right]^2 \left[e^{(\pi L_2/a)} - 4e^{(\pi L_2/2a)} + 3 + \frac{\pi L_2}{a} \right] \quad (2.105)$$

and the low field diffusion coefficient is determined from the Einstein relationship to be

$$D_0 = \frac{kT_0\mu_0}{q} \quad (2.106)$$

The high field diffusion coefficient, D_{hi}, is determined empirically to be about 25 cm²/s [1].

The thermal noise sources associated with the parasitic resistors R_g and R_s can be expressed in terms of a mean square noise voltage given by

$$|\overline{e_m^2}| = 4kT_0R_gB \tag{2.107}$$

$$|\overline{e_f^2}| = 4kT_0R_sB \tag{2.108}$$

Similarly, the real part of the source impedance is given by

$$|\overline{e_s^2}| = 4kT_0R_{sc}B \tag{2.109}$$

The effect of the charging resistor R_i is embedded and included with the gate noise generator i_g.

The overall noise of the FET is determined by adding the contribution from each source. This is accomplished by using the two short-circuit Y parameters

$$Y_{11} = j\omega C_{gs}/(1 + j\omega C_{gs}R_i) \tag{2.110}$$

$$Y_{21} = g_m/(1 + j\omega C_{gs}R_i) \tag{2.111}$$

The noise factor is then expressed as

$$F = 1 + \frac{|\overline{i_{m0} + i_{f0} + i_{g0} + i_{d0}}|^2}{|\overline{i_{sc}}|^2} \tag{2.112}$$

Where i_{m0}, i_{f0}, i_{g0}, and i_{sc} are the noise components in the short-circuited drain-source path produced by the noise generators, e_m, e_f, i_g, i_d, and e_s. Through circuit analysis, the above equation is written in a more common form:

$$F = 1 + (1/R_{sc})(r_n + g_n|Z_{sc} + Z_c|^2) \tag{2.113}$$

where g_n and r_n are referred to as the noise conductance and noise resistance, respectively. The parameter Z_c is called the *correlation impedance*. The parameters are defined as

$$r_n = (R_g + R_s) + K_r\left(\frac{1 + \omega^2C_{gs}^2R_i^2}{g_m}\right) \tag{2.114}$$

$$g_n = K_g\frac{\omega^2C_{gs}^2}{g_m} \tag{2.115}$$

$$Z_c = (R_g + R_s) + K_c/Y_{11} \tag{2.116}$$

The terms K_g, K_r, and K_c are functions of P, R, and C as previously discussed. Pucel defines these K terms as the fundamental noise coefficients and expresses them as

$$K_r = \frac{R(1 - C^2)}{[1 - C(R/P)^{1/2}]^2 + (1 - C^2)R/P} \tag{2.117}$$

$$K_c = \frac{1 - C(R/P)^{1/2}}{[1 - C(R/P)^{1/2}]^2 + (1 - C^2)R/P} \tag{2.118}$$

$$K_g = P\{[1 - C(R/P)^{1/2}]^2 + (1 - C^2)R/P\} \tag{2.119}$$

A minimum noise figure can be obtained by selecting the source impedance to minimize equation (2.113). If we express the optimized source impedance as a complex quantity given by

$$Z_{sc,opt} = R_{sc,opt} + jX_{sc,opt} \tag{2.120}$$

the minimum noise factor can be related to the correlation impedance $Z_c = R_c + jX_c$. The minimum occurs when

$$R_{sc,opt} = [R_c^2 + (r_n/g_n)]^{1/2} \tag{2.121}$$

and

$$X_{sc,opt} = -X_{sc} \tag{2.122}$$

where $R_{sc,opt}$ and $X_{sc,opt}$ are the generator impedances that result in a minimum noise figure.

The minimum noise factor is hence given by

$$F_{min} = 1 + 2g_n(R_c + R_{sc,opt}) \tag{2.123}$$

and the minimum noise figure is given by

$$F_{min} \text{ (dB)} = 10 \log(F_{min}) \tag{2.124}$$

The above equations allow prediction of FET noise parameters as a function of device geometry, material properties, and the bias conditions of the device. While this physically based model is considerably more complex than empirical models, for certain applications, it offers inherent advantages. These include device development, prediction of yield loss due to process fluctuations, and design centering. Such a model also provides a means of examining the theoretical performance of devices that cannot be fabricated currently due to technology limitations.

As an illustration of the ability of the physical model to predict device noise parameters, the dependence of minimum noise figure (F_{min}) on gate width, epi-thickness, and doping density was examined at several drain-source currents [1]. For this illustration, a MESFET with a gate length of 1 μm is chosen. Typical material and device properties for the device fabricated with GaAs are shown in Table 2.4.

Table 2.4
Geometric Parameters and Doping Density Values Chosen for Device Noise Calculations

L = 1 μm	L_{gd} = 2.5 μm
Z = 275 μm	L_s = 100 μm
a = 0.2 μm	L_d = 50 μm
L_{gs} = 1 μm	N_d = 7 $\times$ 10^{16}/cm^{-3}

The results of this simulation are shown in Figures 2.30 through 2.32. As expected, the minimum noise figure decreases monotonically with decreasing gate width (Figure 2.30). However, the drain current, gain, and output power also decrease. Thus, for circuit applications, selecting the optimal gate width for minimum noise constrains other characteristics of the device.

The influence of doping density and epi-thickness is depicted in Figures 2.31 and 2.32, respectively. Better noise performance is generally achieved by reducing these parameters. However, similar to gate width, doping density and epi-thickness affect drain current, gain, and maximum output power. Hence, the minimum noise

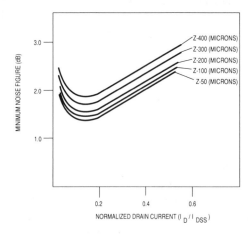

Figure 2.30 Minimum noise figure as a function of normalized drain current and gate width.

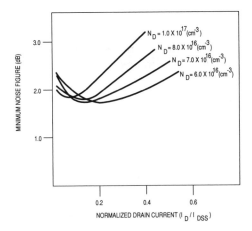

Figure 2.31 Minimum noise figure as a function of normalized drain current and doping density.

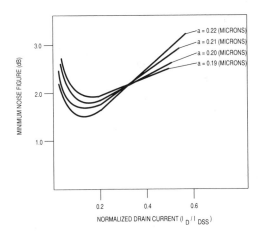

Figure 2.32 Minimum noise figure as a function of normalized drain current and epi-thickness.

figure is achieved at the expense of other important device characteristics and these trade-offs must be considered in achieving low noise devices.

Because HEMT fabrication technology is nearly 20 years less mature than MESFET technology, much less work has been done in the area of physically based noise models for this device. Modifications to the Pucel model to account for HEMT noise mechanisms have been presented [15]. The HEMT noise model shows reasonable agreement with experiments and indicates that HEMT noise is dominated by thermal noise with negligible contribution from high field diffusion noise. This is a significant contrast from the MESFET, where high field diffusion noise often dominates device performance.

2.3 LARGE-SIGNAL MODELS

2.3.1 Empirical MESFET/HEMT Models

Several researchers have developed empirical models that describe the operational characteristics of the GaAs MESFET [16–21]. All of these models are analytical and all are capable of describing the large-signal properties of microwave MESFETs with some success. The MESFET models discussed in this section are available in many of the popular large-signal circuit simulation packages used by microwave engineers. The HEMT models represent more recently derived models, which are not generally available in software packages today. For any given device and application, the optimal choice of models depends on many considerations including model availability, computational efficiency, and prediction accuracy. The accuracy of the models as related to one particular device will be discussed later in this section.

The general empirical modeling approach is to first examine the measured device characteristics and then look for a mathematical function that behaves in the same manner. The mathematical function includes adjustable parameters that, when assigned proper values, will cause the function to approximate the measured data closely. Figure 2.33 is an equivalent circuit corresponding to a typical large-signal model. Analytical large-signal models approximate the nonlinear properties of a device using a unique set of analytical equations. These nonlinear properties can be related to elements of the equivalent circuit. The main nonlinear elements of the circuit of Figure 2.33 are as follows:

(1) the drain-source current I_{ds} controlled by voltages V_{gs} and V_{ds}, and from which the transconductance and output conductance are derived,
(2) the gate-source capacitance C_{gs},
(3) the gate-drain capacitance C_{gd},
(4) the diode D_{gs}, which is important in the modeling of forward-bias gate current (or breakdown under conditions of inverted drain-source bias), and
(5) the diode D_{gd}, which is important in the modeling of drain-gate avalanche current (or forward conduction under conditions of inverted drain-source bias).

The difference between the various large-signal empirical models presented here is the way in which they approximate the current-voltage or capacitance-voltage relationships. To first order, the development of empirical capacitance-voltage rela-

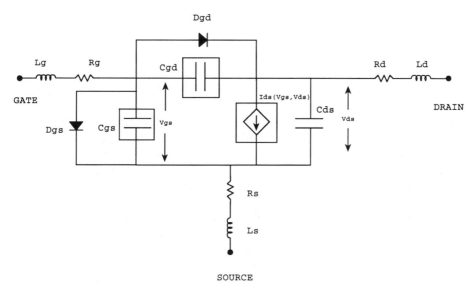

Figure 2.33 An equivalent circuit for a MESFET/HEMT large-signal model.

tionships is independent of the development of the current-voltage relationships. Capacitance-voltage relationships for the MESFET and HEMT are commonly modeled using the simple diffusion capacitance expression to be provided in the discussion of the Curtice model. However, Statz *et al.* have developed capacitance expressions that are more complete for all regions of device operation [14]. These expressions will be provided in the discussion concerning the Statz model. Significantly improved expressions for HEMT capacitance relationships are also presented. The forward conduction in the gate diodes D_{gs} and D_{gd} is typically modeled using the Shockley equation:

$$I_{df} = I_s[e^{V_d/(nV_t)} - 1] \tag{2.125a}$$

where V_d is the voltage across the junction, V_t is the thermal voltage (kT/q), I_s is the reverse saturation current, and n is the ideality factor. The diode parameters I_s and n can be obtained from forward-bias dc measurements described in Sections 3.1 and 4.1.

Breakdown characteristics can also be modeled by utilizing the diode elements. The excess device current due to avalanche breakdown is given to first order by the expression:

$$I_{dr} = I_s[e^{-(V_d + V_{br})/(nV_t)}] \text{ for } V_d < -V_{br} \tag{2.125b}$$

where V_{br} is the device breakdown voltage.

2.3.1.1 Curtice Model

One of the first large-signal MESFET models to be used in a large-signal circuit simulator was proposed by Van Tuyl and Liechti [22] and later simplified by Curtice [16]. The Curtice model consists primarily of the voltage-controlled current source I_{ds}, gate-source capacitance C_{gs}, gate-drain capacitance C_{gd}, drain-source capacitance C_{ds}, and the clamping diode D_{gs}. As originally proposed, the only nonlinear elements in this model are I_{ds} and C_{gs}. The I_{ds} term is a function of the intrinsic (internal to the parasitic resistances and inductances) drain-source and gate-source voltages and a time constant τ, which represents the electron transit time under the gate. The C_{gs} term is considered to be only a function of the intrinsic gate-source voltage. Using I_{ds} and C_{gs} as the only nonlinear elements, however, neglects the highly nonlinear characteristics of C_{gd}. For this reason, most large-signal circuit simulators also include C_{gd} as a nonlinear circuit element that varies only as a function of the intrinsic gate-drain voltage.

The Curtice model describes the drain current with respect to the drain-source and gate-source voltages as

$$I_{ds}(V_{gs}, V_{ds}) = \beta(V_{gs} - V_{TO})^2(1 + \lambda V_{ds}) \tanh(\alpha V_{ds}) \qquad (2.126)$$

where V_{gs} and V_{ds} are the intrinsic terminal voltages and β, V_{TO}, λ, and α are the model parameters. The small signal transconductance and output conductance are derived by differentiating I_{ds} with respect to the gate-source and drain-source voltages, respectively:

$$g_m = \frac{\partial I_{ds}}{\partial V_{gs}} = I_{ds}[2/(V_{gs} - V_{TO})] \qquad (2.127)$$

$$g_{ds} = \frac{\partial I_{ds}}{\partial V_{ds}} = \beta(V_{gs} - V_{TO})^2(1 + \lambda V_{ds})\{\alpha/[\cosh^2(\alpha V_{ds})]\}$$

$$+ \beta(V_{gs} - V_{TO})^2\lambda \tanh(\alpha V_{ds}) \qquad (2.128)$$

If equation (2.126) is separated into three components, we can easily see how the drain current relationship to drain-source voltage and gate-source voltage is modeled. The first component of equation (2.126) is $\beta(V_{gs} - V_{TO})^2$. This is used to model the approximately square-law behavior of the drain current with respect to the gate-source voltage. Figure 2.34 is a graph of $\beta(V_{gs} - V_{TO})^n$ as a function of gate-source voltage with $\beta = 1$, $V_{TO} = 1$, and $n = 1.8$, 2.0, and 2.25. The parameter β is merely a constant with the dimension of A/V^2. As the exponent n is altered, the curvature of the characteristics changes. Other models presented in this section use this exponent as a variable parameter. The next component of equation (2.126) is $1 + \lambda V_{ds}$. The parameter λ is used to model the device output conductance,

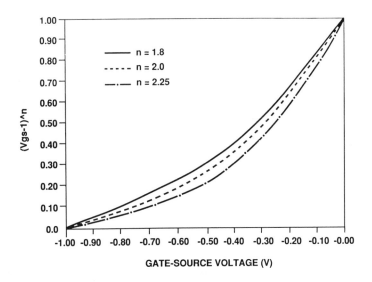

Figure 2.34 A graph of $\beta(V_{gs} - V_{TO})^n$ as a function of gate-source voltage ($\beta = 1$, $V_{TO} = 1$, and $n = 1.8$, 2.0, and 2.25).

which is the slope of the drain-source current with respect to the drain-source voltage. Figure 2.35 is a graph of $1 + \lambda V_{ds}$ *versus* drain-source voltage with values of $\lambda = 0.05$, 0.1, and 0.2. This graph shows that increasing λ causes an increase in slope, which corresponds to an increase in modeled output conductance. The final component of equation (2.126) is $\tanh(\alpha V_{ds})$. The hyperbolic tangent is used because it approximates the $I_{ds} - V_{ds}$ characteristics observed in MESFETs. Figure 2.36 is a graph of $\tanh(\alpha V_{ds})$ *versus* drain-source voltage. This graph illustrates that by increasing α, more rapid current saturation effects can be modeled.

Equation (2.126) does not constitute the complete large-signal model because capacitance-voltage relationships must also be described. The Curtice model uses the capacitance expression derived from first-order semiconductor junction theory applied to a two-terminal Schottky diode structure. The expression used for both the gate-source capacitance and gate-drain capacitance is given by

$$C_{gs,gd} = C_{gs0,gd0} \left(1 - \frac{V_{\text{applied}}}{V_{\text{bi}}} \right)^{-1/2} \tag{2.129}$$

where V_{bi} is the built-in potential of the Schottky gate, V_{applied} is the intrinsic gate-source or gate-drain voltage, and $C_{gs0,gd0}$ is the zero-bias gate-source or gate-drain capacitance. Note that equation (2.129) does not include drain-source voltage dependence, which may be important for many applications. Some models also use

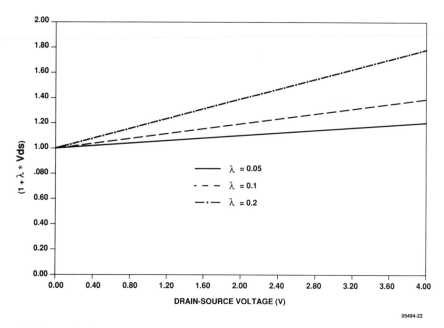

Figure 2.35 A graph of $1 + \lambda V_{ds}$ as a function of drain-source voltage for $\lambda = 0.05, 0.1,$ and 0.2.

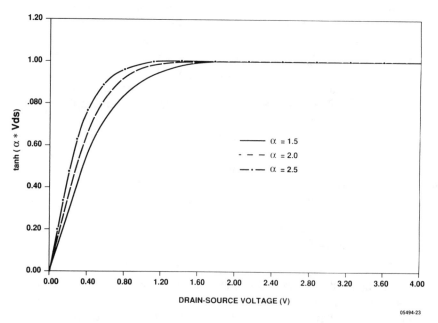

Figure 2.36 A graph of $\tanh(\alpha V_{ds})$ as a function of drain-source voltage for values of $\alpha = 1.5, 2.0,$ and 2.5.

the exponential [equal to $-\frac{1}{2}$ in equation (2.129)] as a parameter in order to model the capacitance-voltage relationship more closely. Other expressions are also sometimes utilized to represent the capacitance when the device is forward biased.

In addition to current-voltage and capacitance-voltage relationships, the Curtice model also accounts for the time delay between applied gate voltage fluctuations and observed drain current changes. The resulting time delay can be modeled by assuming the current source is of the form:

$$I(V) - \tau \frac{dI(V)}{dt} \tag{2.130}$$

where the derivative can be evaluated as

$$\frac{dI(V)}{dt} = \left[\frac{\partial I(V)}{\partial V_{gs}} \right] \frac{dV_{gs}}{dt} \tag{2.131}$$

Although this technique is easily expressed mathematically, many large-signal simulators do not include this modification due to the resulting increase in required simulation time.

2.3.1.2 Statz (Raytheon) Model

The researchers H. Statz, P. Newman, I. Smith, R. Pucel, and H. Haus developed a model based on equation (2.126). Their research [17] indicates that the square-law approximation of drain-source current as a function of gate-source voltage is only valid for small $V_{gs} - V_{TO}$ values, where V_{TO} is the pinch-off potential. The drain-source current I_{ds} becomes almost linear for larger values of $V_{gs} - V_{TO}$. To model this behavior, the original square-law expression:

$$I_{ds} \sim \beta(V_{gs} - V_{TO})^2 \tag{2.132}$$

is modified. The new expression is

$$I_{ds} \sim \frac{\beta(V_{gs} - V_{TO})^2}{1 + b(V_{gs} - V_{TO})} \tag{2.133}$$

For small values of $V_{gs} - V_{TO}$, the new expression is quadratic, while for large values, it behaves in a linear fashion in $V_{gs} - V_{TO}$. Figure 2.37 is a graphical comparison of equations (2.132) and (2.133) as a function of gate-source voltage. For equation (2.132), $\beta = 1$ and $V_{TO} = 1$; for equation (2.133), $\beta = 1.25$, $V_{TO} = 1$, and $b = 0.25$.

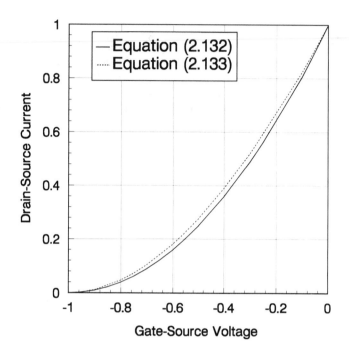

Figure 2.37 A graphical comparison of equations (2.132) and (2.133) as a function of a gate-source voltage. For equation (2.132), $\beta = 1$ and $V_{TO} = 1$. For equation (2.133), $\beta = 1.25$, $V_{TO} = 1$, and $b = 0.25$.

Finally, the hyperbolic tangent function tends to consume significant amounts of computer time and, therefore, is modified using a polynomial of the form:

$$1 - [1 - (\alpha V_{ds}/n)]^n$$

with $n = 3$. Although this representation does not look similar to the tanh representation, it is simply a truncated series portrayal of the hyperbolic tangent. In the saturation region, the polynomial is replaced by unity. These modifications lead to a new form for the drain current and its derivatives. For $0 < V_{ds} < 3/\alpha$,

$$I_{ds} = \frac{\beta(V_{gs} - V_{TO})^2}{1 + b(V_{gs} - V_{TO})} \left[1 - \left(1 - \frac{\alpha V_{ds}}{3} \right)^3 \right] (1 + \lambda V_{ds}) \tag{2.134a}$$

$$g_m = \left[1 - \left(1 - \frac{\alpha V_{ds}}{3} \right)^3 \right] (1 + \lambda V_{ds}) \left[\frac{2\beta(V_{gs} - V_{TO})}{1 + b(V_{gs} - V_{TO})} \right]$$

$$- \frac{b\beta(V_{gs} - V_{TO})^2}{[1 + b(V_{gs} - V_{TO})]^2} \tag{2.134b}$$

$$g_{ds} = \left[1 - \left(1 - \frac{\alpha V_{ds}}{3}\right)^3\right]\lambda + \alpha(1 + \lambda V_{ds})\left(1 - \frac{\alpha V_{ds}}{3}\right)^2$$

$$\cdot \frac{\beta(V_{gs} - V_{TO})^2}{1 + b(V_{gs} - V_{TO})} \tag{2.134c}$$

For $V_{ds} \geq 3/\alpha$,

$$I_{ds} = \frac{\beta(V_{gs} - V_{TO})^2(1 + \lambda V_{ds})}{1 + b(V_{gs} - V_{TO})} \tag{2.135a}$$

$$g_m = \frac{[1 + b(V_{gs} - V_{TO})]2\beta(V_{gs} - V_{TO}) - b\beta(V_{gs} - V_{TO})^2}{[1 + b(V_{gs} - V_{TO})]^2}$$

$$\cdot (1 + \lambda V_{ds}) \tag{2.135b}$$

$$g_{ds} = \frac{\lambda\beta(V_{gs} - V_{TO})^2}{1 + b(V_{gs} - V_{TO})} \tag{2.135c}$$

New capacitance expressions replace the simple junction capacitance of the Curtice model. The simple Curtice model junction capacitance expression fails to predict accurate capacitance values for low drain-source voltages. The model also fails when the drain and source bias levels reverse. This is due, in part, to the lack of dependence on the drain-source voltage. The Statz model expressions for capacitance are developed to correct these shortfalls. These capacitance-voltage expressions are:

$$C_{gs} = (C_{gs0}K2K1/(1 - V_n/V_{bi})^{1/2} + C_{gd0}K3) \tag{2.136}$$

$$C_{gd} = (C_{gs0}K3K1/(1 - V_n/V_{bi})^{1/2} + C_{gd0}K2) \tag{2.137}$$

where

$$V_{max} = 0.5$$
$$\delta = 0.2$$
$$K1 = \{1 + (V_e - V_{TO})/[(V_e - V_{TO})^2 + \delta^2]^{1/2}\}/2$$
$$K2 = \{1 + (V_{gs} - V_{gd})/[(V_{gs} - V_{gd})^2 + (1/\alpha)^2]^{1/2}\}/2$$
$$K3 = \{1 - (V_{gs} - V_{gd})/[(V_{gs} - V_{gd})^2 + (1/\alpha)^2]^{1/2}\}/2$$
$$V_e = \{V_{gs} + V_{gd} + [(V_{gs} - V_{gd})^2 + (1/\alpha)^2]^{1/2}\}/2$$
$$\text{If: } \{V_e + V_{TO} + [(V_e - V_{TO})^2 + \delta^2]^{1/2}\}/2 < V_{max}$$
$$V_n = \{V_e + V_{TO} + [(V_e - V_{TO})^2 + \delta^2]^{1/2}\}/2$$
$$\text{else: } V_n = V_{max}$$

The voltages V_{gs} and V_{gd} are intrinsic device voltages. The adjustable model parameters are C_{gs0}, C_{gd0}, and α, although α is usually determined by consideration of equations (2.134) and (2.135). Figures 2.38 and 2.39 are graphs of the modeled gate-source and gate-drain capacitances, respectively, *versus* drain-source voltage. Notice that as the drain-source voltage becomes negative, the gate-source capacitance approaches the gate-drain capacitance. When the device is pinched off, the gate-source capacitance descends to a very small value. Notice that for large negative values of drain-source voltage, the gate-source capacitance is independent of gate-source voltage. Figure 2.40 illustrates the behavior of gate-source capacitance using equation (2.136) as a function of gate-source voltage for various values of drain-source voltage. For drain-source voltages in the normal operating range, equation (2.136) approximates a diode-like capacitance model. As V_{gs} approaches the pinch-off voltage (in this case, $V_{TO} = 1.0$ V) the gate-source capacitance falls rapidly to zero.

The Statz capacitance expressions are more sophisticated mathematically than the simple diffusion capacitance model of equation (2.129). The Statz expressions consider voltage dependence of both drain-source and gate-source potentials and qualitatively resemble the measured characteristics of the device as presented

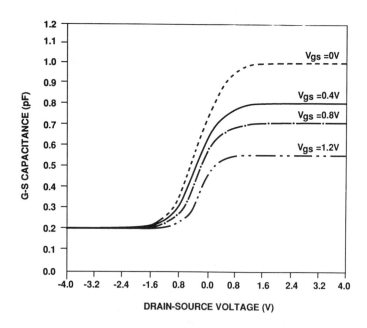

Figure 2.38 Modeled gate-source capacitance using equation (2.136) as a function of drain-source voltage for gate-source voltages of 0.0, -0.4, -0.8, and -1.2 V. ($C_{gs0} = 1$, $C_{gd0} = 0.2$, $\delta = 0.2$, $V_{max} = 0.5$, $V_{TO} = 2.0$, and $\alpha = 2.0$.)

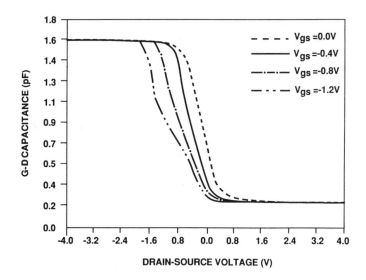

Figure 2.39 Modeled gate-drain capacitance using equation (2.137) as a function of drain-source voltage for gate-source voltages of 0.0, -0.4, -0.8, and -1.2 V. ($C_{gs0} = 1$, $C_{gd0} = 0.2$, $\delta = 0.2$, $V_{max} = 0.5$, $V_{TO} = 2.0$, and $\alpha = 2.0$.)

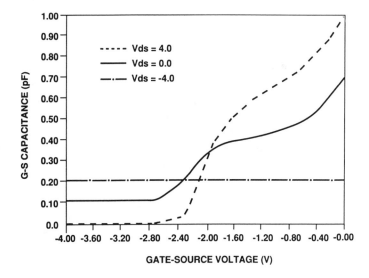

Figure 2.40 Modeled gate-source capacitance using equation (2.136) as a function of gate-source voltage for drain-source voltages of 4.0, 0.0, and -4.0 V. ($C_{gs0} = 1$, $C_{gd0} = 0.2$, $\delta = 0.2$, $V_{max} = 0.5$, $V_{TO} = 2.0$, and $\alpha = 2.0$.)

in Chapter 1. Despite these facts, the expressions often fail to agree quantitatively with measured characteristics of some devices.

2.3.1.3 Materka-Kacprzak Model

The Materka-Kacprzak model [18] is based on the FET model by Taki [23], however the pinch-off potential is modified to account for a drain-source bias dependence. The drain current and its derivatives are given by

$$I_{ds} = I_{dss}\left(1 - \frac{V_{gs}}{V_T}\right)^2 \tanh[\alpha V_{ds}/(V_{gs} - V_T)] \tag{2.138}$$

$$g_m = \tanh[\alpha V_{ds}/(V_{gs} - V_T)]\left[\frac{-2I_{dss}}{V_T}\left(1 - \frac{V_{gs}}{V_T}\right)\right]$$

$$- I_{dss}(1 - V_{gs}/V_T)^2 \operatorname{sech}^2[\alpha V_{ds}/(V_{gs} - V_T)]\left[\frac{\alpha V_{ds}}{(V_{gs} - V_T)^2}\right] \tag{2.139}$$

$$g_{ds} = 2I_{dss}(1 - V_{gs}/V_T)\frac{\gamma V_{gs}}{V_T^2}\tanh[\alpha V_{ds}/(V_{gs} - V_T)]$$

$$+ I_{dss}(1 - V_{gs}/V_T)^2 \operatorname{sech}^2[\alpha V_{ds}/(V_{gs} - V_T)]$$

$$\cdot \frac{\alpha(V_{gs} - V_T) + \alpha\gamma V_{ds}}{(V_{gs} - V_T)^2} \tag{2.140}$$

$$V_T = V_{TO} + \gamma V_{ds} \tag{2.141}$$

The parameter I_{dss} is the saturation current at zero gate-source voltage. The modified pinch-off potential is represented by the parameter V_T with V_{TO} representing the pinch-off potential of an ideal square-law device. In addition, α and γ are two empirical parameters. The parameter α is used as it is in the Curtice model, while γ describes the effective pinch-off when combined with V_{ds}. The four parameters I_{dss}, α, γ, and V_{TO} are used to fit the measured data. The nonlinear capacitances C_{gs} and C_{gd} are given by equation (2.129). The current for diode D_{gs} is given by

$$I_{df} = I_s[e^{(\alpha_s V_{gs})} - 1] \tag{2.142}$$

where the model parameters are I_s and α_s and V_{gs} is the intrinsic gate-source voltage. The current for diode D_{dg} is given by

$$I_{dr} = I_{sr}[e^{(\alpha_{sr} V_{dg})} - 1] \tag{2.143}$$

where I_{sr} and α_{sr} are the model parameters and V_{dg} is the intrinsic drain-gate voltage.

2.3.1.4 TriQuint's Own Model (TOM)

TOM is a variation of the Statz model with three important modifications [19]. The first change is in the square-law term $(V_{gs} - V_{TO})^2$. Actual devices do not exhibit exact quadratic behavior; to account for this, the exponent in the square-law term is changed to an adjustable model parameter. Second, the pinch-off potential is modified to account for a drain-source voltage dependence. The new pinch-off potential is given by equation (2.141) with γ replaced by $-\gamma$, where V_{TO} is the ideal device pinch-off potential and γV_{ds} is the effective pinch-off potential displacement. The final change deals with the models' abilities to match drain conductance. The Statz model can only fit drain conductance for a small range of drain current. When drain current increases out of this neighborhood, the Statz model tends to predict conductance values that are too large. The decrease of the device drain conductance, at low gate-source bias, is modeled by a parameter δ. The modified set of equations for drain current I_{ds} and its derivatives follow. For $0 < V_{ds} < 3/\alpha$,

$$I_{ds} = I_d \{1 - [1 - (\alpha V_{ds})/3]^3\} \tag{2.144a}$$

$$g_m = \{1 - [1 - (\alpha V_{ds})/3]^3\} \left[\frac{Q\beta(V_{gs} - V_T)}{1 + \delta V_{ds}I_{ds0}} - \frac{I_{ds0}(Q\delta V_{ds}\beta)(V_{gs} - V_T)}{(1 + \delta V_{ds}I_{ds0})^2} \right] \tag{2.144b}$$

$$g_{ds} = \frac{+Q\beta\gamma(V_{gs} - V_T)(1 + \delta V_{ds}I_{ds0}) - I_{ds0}\{\delta[I_{ds0} + Q\beta\gamma V_{ds}(V_{gs} - V_T)]\}}{(1 + \delta V_{ds}I_{ds0})^2}$$
$$\cdot \{1 - [1 - (\alpha V_{ds})/3]^3\} + I_{ds}\{\alpha[1 - (\alpha V_{ds})/3]^2\} \tag{2.144c}$$

where

$$I_d = \frac{I_{ds0}}{1 + \delta V_{ds}I_{ds0}}$$

$$I_{ds0} = \beta(V_{gs} - V_T)^Q$$

For $V_{ds} \geq 3/\alpha$,

$$I_{ds} = I_d \tag{2.145a}$$

$$g_m = \frac{Q\beta(V_{gs} - V_T)}{1 + \delta V_{ds}I_{ds0}} - \frac{QI_{ds0}\delta V_{ds}\beta(V_{gs} - V_T)}{(1 + \delta V_{ds}I_{ds0})^2} \tag{2.145b}$$

$$g_{ds} = \frac{+Q\beta\gamma(V_{gs} - V_T)(1 + \delta V_{ds}I_{ds0})}{(1 + \delta V_{ds}I_{ds0})^2}$$
$$- \frac{I_{ds0}\{\delta[I_{ds0} + Q\beta\gamma V_{ds}(V_{gs} - V_T)]\}}{(1 + \delta V_{ds}I_{ds0})^2} \tag{2.145c}$$

The new parameters γ and δ replace λ and b in the Statz model. These modified equations, in conjunction with the gate diode and gate capacitance expressions from the original Statz model, constitute the full large-signal model.

2.3.1.5 Advanced Curtice Model (Meta-Software)

The advanced Curtice model is a modified version of the original Curtice model [20]. As in previous models, the pinch-off potential is modified to account for a drain-source voltage dependence. The revised pinch-off potential is given by equation (2.141), where γ is the parameter used to adjust the effective pinch-off voltage. The transconductance parameter β is also modified to fit the transconductance in accordance with actual device behavior. The new effective β is expressed as

$$\beta_{\text{eff}} = \frac{\beta}{(1 + \mu_{\text{crit}} V_{gst})} \tag{2.146}$$

where $V_{gst} = V_{gs} - V_T$, and μ_{crit} is the parameter representing the critical field for mobility degradation in V/cm. In saturation, the transconductance can be approximated as $2\beta(V_{gs} - V_{TO})$. Figure 2.41 represents the effect of β_{eff} on the prediction of transconductance with respect to the gate-source voltage. Notice that as the gate-source voltage is decreased from the pinch-off voltage, transconductance tends to

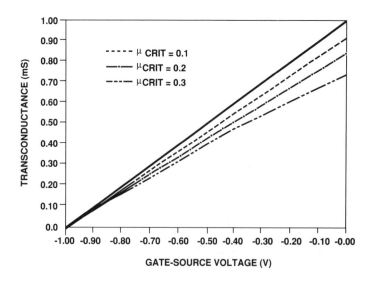

Figure 2.41 A graphical comparison of modeled transconductance using equation (2.146) illustrating the effect of β_{eff} on transconductance.

roll over. This is a closer approximation to the real device transconductance. Without this correction to β, the transconductance is modeled as a linear function of the gate-source voltage. The final modification is similar to previous models in that the square-law term is allowed to vary according to an exponential parameter (VGEXP). The drain-source current and its derivatives are expressed as [20]:

$$I_{ds} = \beta_{\mathrm{eff}}(V_{gst}^{\mathrm{VGEXP}})(1 + \lambda V_{ds})\tanh(\alpha V_{ds}) \qquad (2.147)$$

$$g_m = I_{ds}[(\mathrm{VGEXP}/V_{gst}) - \mu_{\mathrm{crit}}/(1 + \mu_{\mathrm{crit}}V_{gst})] \qquad (2.148)$$

$$g_{ds} = \beta_{\mathrm{eff}}(V_{gst}^{\mathrm{VGEXP}})(1 + \lambda V_{ds})[\alpha/\cosh^2(\alpha V_{ds})] \qquad (2.149)$$
$$+ \beta_{\mathrm{eff}}(V_{gst}^{\mathrm{VGEXP}})\lambda\tanh(\alpha V_{ds}) - g_m\gamma$$

2.3.1.6   Curtice-Ettenberg Model

Curtice and Ettenberg [21] altered the original Curtice model in order to produce a closer fit to the relationship between the drain source current as a function of the gate-source voltage. In the original model, this relationship was assumed to behave in a square-law manner. An alternative solution to allowing the exponent of the $(V_{gs} - V_T)$ term to take on values other than 2 is to utilize a higher order polynomial in gate-source voltage. A cubic approximation to this relationship, for example, provides the modeler with additional freedom in matching the device characteristics. With the cubic approximation, the new equations for the drain-source current and its derivatives are [21]:

$$I_{ds} = (A_0 + A_1 v1 + A_2 v1^2 + A_3 v1^3)\tanh(\gamma V_{ds}) \qquad (2.150)$$

$$g_m = \tanh(\gamma V_{ds})(A_1 v2 + 2A_2 v1 v2 + 3A_3 v1^2 v2) \qquad (2.151)$$

$$g_{ds} = (A_0 + A_1 v1 + A_2 v1^2 + A_3 v1^3)\operatorname{sech}^2(\gamma V_{ds})\gamma \qquad (2.152)$$
$$- \beta V_{gs}(A_1 + 2A_2 v1 + 3A_3 v1^2)\tanh(\gamma V_{ds}) + 1/V_{ds0}$$

with

$$v1 = V_{gs}[1 + \beta(V_{ds0} - V_{ds})]$$
$$v2 = [1 + \beta(V_{ds0} - V_{ds})]$$

The voltage $v1$ is used to model the occurrence of pinch-off voltage increase with drain-source voltage. The parameter β is now used to control the change in pinch-off voltage; V_{ds0} is the drain-source voltage (in saturation) at which the A_i coefficients are evaluated. Gate diode and gate capacitance characteristics, as discussed earlier, are included to complete the full large-signal model.

2.3.1.7 Procedure for Developing HEMT Models

Although HEMT devices have already exhibited superior high frequency and low noise performance over competing MESFET technology, much less work has been done to produce accurate large-signal HEMT models that are compatible with large-signal circuit simulator routines.

As previously described, GaAs MESFET models successfully predict large-signal performance by predicting the voltage dependence of device characteristics. Of the nonlinearities involved in device behavior, transconductance is most critical to the accurate prediction of many important large-signal effects. As pointed out in Chapter 1, however, the differences between the transconductance characteristics of HEMTs and MESFETs is one of the most significant distinctions between the device's electrical behavior.

To accomplish an improved prediction of transconductance for the HEMT, the analytical transconductance expression of existing FET models is modified [24]:

$$g_m = g_{mFET} - f(V_{ds})\xi(V_{gs} - V_{pf})^{\psi} \tag{2.153}$$

where g_{mFET} is the MESFET model transconductance; $f(V_{ds})$ is a nonlinear separable function of V_{ds}; V_{pf} is a voltage at which transconductance begins to degrade; and ξ and ψ are empirical transconductance degradation parameters. The motivation for this modification is that transconductance begins to degrade for voltages greater than some gate-source voltage level, V_{pf}. This is accounted for by the second term of equation (2.153). The rate of degradation is determined by the exponent ψ in the equation.

The transconductance as modified by this equation is then integrated with respect to V_{gs} to derive the I-V characteristics of the model

$$I_{ds} = \int g_m dV_{gs} \tag{2.154}$$

which can also be expressed in terms of a standard MESFET current with a modification term introduced by the transconductance degradation term. An expression for output conductance is then derived by taking the derivative of the I-V equation with respect to V_{ds}, yielding,

$$g_{ds} = g_{dsFET} - g_{dshpf} \tag{2.155}$$

where g_{dsFET} is the MESFET model output conductance and g_{dshpf} is a term that results from the HEMT transconductance degradation terms.

2.3.1.8 HEMT Model (Curtice)

To apply the procedure to the Curtice model [16], we begin with the Curtice model *I-V* expression:

$$I_{dsFET} = \begin{cases} 0 & \text{for } V_{gs} \leq V_{TO} \\ \beta(V_{gs} - V_{TO})^2 \tanh(\alpha V_{ds})(1 + \lambda V_{ds}) & \text{for } V_{gs} > V_{TO} \end{cases} \tag{2.156}$$

Transconductance for this model is then found by taking the derivative of (2.156) with respect to gate-source voltage to yield

$$g_{mFET} = \begin{cases} 0 & \text{for } V_{gs} \leq V_{TO} \\ 2\beta(V_{gs} - V_{TO})f(V_{ds}) & \text{for } V_{gs} > V_{TO} \end{cases} \tag{2.157}$$

where $f(V_{ds}) = \tanh(\alpha V_{ds})(1 + \lambda V_{ds})$ is the nonlinear separable function of V_{ds}. To add the HEMT correction terms, a third region of operation is defined for the device. The modified transconductance is then expressed as

$$g_m = \begin{cases} g_{mFET} & \text{for } V_{gs} \leq V_{pf} \\ g_{mFET} - f(V_{ds})\xi(V_{gs} - V_{pf})^\psi & \text{for } V_{gs} > V_{pf} \end{cases} \tag{2.158}$$

Carrying through the integration prescribed by equation (2.154) results in a current-voltage expression given by

$$I_{ds} = \begin{cases} I_{dsFET} & \text{for } V_{gs} \leq V_{pf} \\ I_{dsFET} - \dfrac{\xi}{\psi + 1}(V_{gs} - V_{pf})^{\psi+1}f(V_{ds}) & \text{for } V_{gs} > V_{pf} \end{cases} \tag{2.159}$$

Finally, the output conductance is obtained from equation (1.2):

$$g_{ds} = \begin{cases} g_{dsFET} & \text{for } V_{gs} \leq V_{pf} \\ g_{dsFET} - \dfrac{\xi}{\psi + 1}(V_{gs} - V_{pf})^{\psi+1}\dfrac{df(V_{ds})}{dV_{ds}} & \text{for } V_{gs} > V_{pf} \end{cases} \tag{2.160}$$

In addition to transconductance, the capacitance characteristics of HEMTs also differ significantly from those observed in MESFETs. This is illustrated in the characteristics presented in Chapter 1. An accurate HEMT model needs to account for these differences. Capacitance-voltage relationships that utilize Meyer's empirical MOSFET capacitance model [25], as well as specifically developed HEMT terms, can accurately describe HEMT C-V characteristics. The gate-source and gate-drain capacitances are calculated by using the following expressions:

$$C_{gs} = \begin{cases} \frac{2}{3}\left[1 - \frac{(V_{dss} - V_{ds})^2}{(2V_{dss} - V_{ds})^2}\right] C_G + C_{GS0} & \text{for } V_{ds} < V_{dss} \\ \frac{2}{3}C_G + C_{GS0} & \text{for } V_{ds} \geq V_{dss} \end{cases} \tag{2.161}$$

$$C_{gd} = \begin{cases} \frac{2}{3}\left[1 - \frac{V_{dss}^2}{(2V_{dss} - V_{ds})^2}\right] C_G + C_{GD0} & \text{for } V_{ds} < V_{dss} \\ C_{GD0} & \text{for } V_{ds} \geq V_{dss} \end{cases} \tag{2.162}$$

$$C_G = \begin{cases} C_{m0}(V_{gs} - V_{TO})^{1/\chi} & \text{for } V_{gs} > V_{TO} \\ 0 & \text{for } V_{gs} \leq V_{TO} \end{cases} \tag{2.163}$$

$$V_{dss} = \begin{cases} V_{ds0}\left(1 - \frac{V_{gs}}{V_{TO}}\right) & \text{for } V_{gs} > V_{TO} \\ 0 & \text{for } V_{gs} \leq V_{TO} \end{cases} \tag{2.164}$$

The HEMT capacitance model parameters are:

V_{TO} = device threshold voltage (determined from I-V relationships for model),

V_{ds0} = drain-source voltage where current saturation occurs with zero gate bias,

C_{GS0} = gate-source fixed fringing capacitance (i.e., minimum value of C_{gs}),

C_{GD0} = gate-drain fixed fringing capacitance (i.e., minimum value of C_{gd}),

C_{m0} = empirical total gate electrode capacitance parameter, and

χ = empirical gate electrode coefficient parameter

These six parameters describe the gate-source and gate-drain capacitance. Figures 2.42(a) and (b) present a comparison between measured HEMT capacitance and the predictions resulting from equations (2.161) through (2.164). The parameter V_{TO} is required by most large-signal models to describe I-V behavior, so it does not represent an additional parameter. For many devices, C_{GS0} and C_{GD0} can be set equal. These parameters are easily determined from measured C-V data for a device or can be extracted using the techniques described in Chapter 4.

Equations (2.158) through (2.160), along with capacitance-voltage expressions, provide the basic formula required for a model to be incorporated into a large-signal circuit simulator. When applied to the other MESFET models, the prescribed derivations are more tedious but still straightforward.

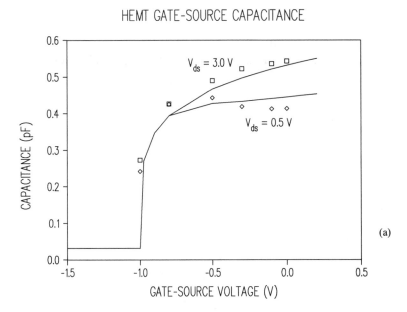

HEMT GATE-SOURCE CAPACITANCE

(a)

CAPACITANCE (pF)

GATE-SOURCE VOLTAGE (V)

$V_{ds} = 3.0$ V

$V_{ds} = 0.5$ V

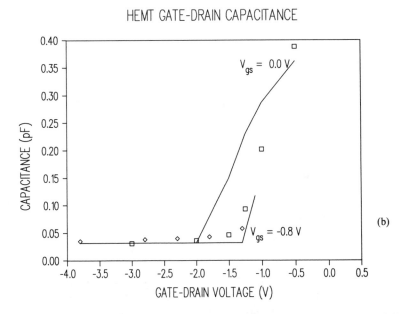

HEMT GATE-DRAIN CAPACITANCE

(b)

CAPACITANCE (pF)

GATE-DRAIN VOLTAGE (V)

$V_{gs} = 0.0$ V

$V_{gs} = -0.8$ V

Figure 2.42 A comparison of measured and modeled capacitance for the HEMT using equations (2.161) through (2.164). The points are measured data while the solid lines are modeled: (a) gate-source capacitance; (b) gate-drain capacitance.

2.3.1.9 HEMT Materka-Kacprzak Model

The Materka-Kacprzak MESFET model [18] is modified using the previously described technique to obtain a new model for HEMTs. The new device equation for $V_{gs} < V_{pf}$ are:

$$I_{ds} = I_{dss}\left(1 - \frac{V_{gs}}{V_T}\right)^2 \tanh[\alpha V_{ds}/(V_{gs} - V_T)] \tag{2.165a}$$

$$g_m = \tanh[\alpha V_{ds}/(V_{gs} - V_T)]\left[\frac{-2I_{dss}}{V_T}\left(1 - \frac{V_{gs}}{V_T}\right)\right]$$
$$- I_{dss}(1 - V_{gs}/V_T)^2 \operatorname{sech}^2[\alpha V_{ds}/(V_{gs} - V_T)]$$
$$\cdot \frac{\alpha V_{ds}}{(V_{gs} - V_T)^2} \tag{2.165b}$$

$$g_{ds} = 2I_{dss}(1 - V_{gs}/V_T)\frac{\gamma V_{gs}}{V_T^2}\tanh[\alpha V_{ds}/(V_{gs} - V_T)]$$
$$+ I_{dss}(1 - V_{gs}/V_T)^2 \operatorname{sech}^2[\alpha V_{ds}/(V_{gs} - V_T)]$$
$$\cdot \frac{\alpha(V_{gs} - V_T) + \alpha\gamma V_{ds}}{(V_{gs} - V_T)^2} \tag{2.165c}$$

For $V_{gs} \geq V_{pf}$

$$I_{ds} = I_{dss}(1 - V_{gs}/V_T)^2 - \xi(V_{gs} - V_{pf})^\psi \tanh[\alpha V_{ds}/(V_{gs} - V_T)] \tag{2.166a}$$
$$g_m = -I_{dss}[(1 - V_{gs}/V_T)^2 - \xi(V_{gs} - V_{pf})^\psi]\{1 - \tanh^2[\alpha V_{ds}/(V_{gs} - V_T)]\}$$
$$\cdot (\alpha V_{ds}/(V_{gs} - V_T) - I_{dss}\tanh(\alpha V_{ds}/(V_{gs} - V_T)(2/V_T(1 - V_{gs}/V_T)$$
$$+ [\psi\xi(V_{gs} - V_{pf})^{\psi-1}] \tag{2.166b}$$
$$g_{ds} = [\alpha/(V_{gs} - V_T)] + [(\alpha\gamma V_{ds})/(V_{gs} - V_T)]\{1 - \tanh^2[\alpha V_{ds}/(V_{gs} - V_T)]\}$$
$$\cdot I_{dss}[(1 - V_{gs}/V_T)^2 - \xi(V_{gs} - V_{pf})^\psi] + 2I_{dss}(1 - V_{gs}/V_T)$$
$$\cdot (\alpha V_{gs}/V_T^2)\tanh^2[\alpha V_{ds}/(V_{gs} - V_T)] \tag{2.166c}$$

with $V_T = V_{TO} + \gamma V_{ds}$. The parameter I_{dss} is the zero gate-source voltage saturation current, V_T is the modified pinch-off potential with V_{TO} representing the pinch-off potential of an ideal square-law device. The capacitance-voltage relationships of (2.161) through (2.164) can be used with this model.

2.3.1.10   HEMT Advanced Curtice Model

The prescribed method is also used to derive a new HEMT model based on the MESFET advanced Curtice model [20]. The new device equations for $V_{gs} < V_{pf}$ are:

$$I_{ds} = \beta_{\text{eff}}(V_{gst}^{\text{VGEXP}})(1 + \lambda V_{ds})\tanh(\alpha V_{ds}) \tag{2.167a}$$

$$g_m = I_{ds}[(\text{VGEXP}/V_{gst}) - \mu_{\text{crit}}/(1 + \mu_{\text{crit}}V_{gst})] \tag{2.167b}$$

$$g_{ds} = \beta_{\text{eff}}(V_{gst}^{\text{VGEXP}})(1 + \lambda V_{ds})[\alpha/\cosh^2(\alpha V_{ds})] \tag{2.167c}$$
$$+ \beta_{\text{eff}}(V_{gst}^{\text{VGEXP}})\lambda\tanh(\alpha V_{ds}) - g_m\gamma$$

For $V_{gs} \geq V_{pf}$,

$$I_{ds} = \beta_{\textit{eff}}V_{gst}^{\text{VGEXP}}(1 + \lambda V_{ds})\tanh(\alpha V_{ds}) - \beta_{\text{eff}}\xi(V_{gs} - V_{pf})^{\psi} \tag{2.168a}$$
$$\cdot (1 + \lambda V_{ds})\tanh(\alpha V_{ds})$$

$$g_m = \beta_{\text{eff}}V_{gst}^{\text{VGEXP}}(1 + \lambda V_{ds})\tanh(\alpha V_{ds})[\text{VGEXP}/V_{gst}$$
$$- \mu_{\text{crit}}/(1 + \mu_{\text{crit}}V_{gst})] - \beta_{\text{eff}}\xi(V_{gs} - V_{pf})^{\psi}(1 + \lambda V_{ds})$$
$$\cdot \tanh(\alpha V_{ds})[\psi/(V_{gs} - V_{pf})] - [\mu_{\text{crit}}/(1 + \mu_{\text{crit}}V_{gst})] \tag{2.168b}$$

$$g_{ds} = \beta_{\text{eff}}(V_{gst}^{\text{VGEXP}})(1 + \lambda V_{ds})[\alpha/\cosh^2(\alpha V_{ds})]$$
$$+ \beta_{\text{eff}}(V_{gst}^{\text{VGEXP}})\lambda\tanh(\alpha V_{ds}) - g_m\gamma - \beta_{\text{eff}}\xi(V_{gs} - V_{pf})^{\psi}$$
$$\cdot (1 + \lambda V_{ds})\tanh(\alpha V_{ds})[(\mu_{\text{crit}}\gamma)/(1 + \mu_{\text{crit}}V_{gst})]$$
$$+ [\lambda/(1 + \lambda V_{ds})] + \alpha(1 - \tanh^2(\alpha V_{ds}))/\tanh(\alpha V_{ds}) \tag{2.168c}$$

The capacitance-voltage relationships expressed by equations (2.161) through (2.164) may be used to complete this model.

2.3.1.11   MESFET/HEMT Model Comparisons

Choosing the optimum large-signal model can be a problem. Factors to consider in making such a decision include availability of the model in circuit simulation routines, the model requirements for the particular application, the computational efficiency of model evaluation and the accuracy of the model predictions. Once a model is chosen, the process of determining ideal model parameters and evaluating the ultimate potential of the models is also difficult. Problems related to model accuracy and determination of model parameters can be overcome by evaluating the ability of a large-signal model to match measured device behavior. A model's ability to match measured data can be benchmarked by determining the minimum error between measured and modeled data. This minimum error value can then be used to establish which model best describes measured data.

All of the large-signal MESFET models presented above, as well as the three HEMT models, have been compared in this manner. The comparisons presented are for only one particular MESFET or HEMT. A number of different devices have been compared with this method, however. These devices have included both MESFETs and HEMTs with gate lengths ranging from 0.25 to 1.0 μm. Low noise, medium power, and high power devices have been investigated. Devices from more than ten different foundries were examined. For every device considered, the relative ranking of the models has remained the same as that illustrated in the figures. The error between modeled and measured data for each model is quantified and minimized using a variation of the Levenberg-Marquardt method as described in Section 4.3. This minimum obtainable error is used as the basis for comparing the models. Figure 2.43 is a bar graph illustrating the relative performance of each MESFET model for both global (linear through saturation regions of device operation) and local (saturation region of operation) optimization. The general trend observed for this particular device (as well as others considered) from best match to worse is the advanced Curtice model (model 2), the TriQuint model (model 6),

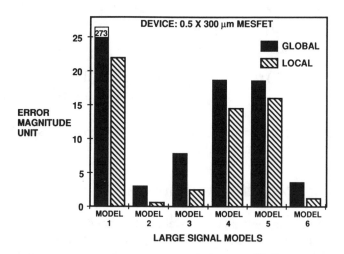

Figure 2.43 The error magnitude unit (EMU) for the various models applied to a 0.5×300-μm MESFET. This illustrates the comparison of minimum obtainable error between measured and modeled data for six large-signal FET models. Model 1 is the Curtice-Ettenberg Model [21]; model 2 is the advanced Curtice model [20]; model 3 is the Materka-Kacprzak model [18]; model 4 is the Statz model [17]; model 5 is the Curtice model [16]; and model 6 is the TriQuint model [19].

the Materka-Kacprzak model (model 3), the Statz model (model 4), the Curtice model (model 5), and the Curtice-Ettenberg model (model 1). The Curtice-Ettenberg model was developed for use with a specific parameter extraction algorithm that does not consider global optimization of all parameters at one time. This model does not lend itself well to the global extraction process used here. The final parameter and EMU values obtained for this model (using the prescribed approach) are extremely sensitive to the initial guess of the parameter values. For some measured data sets, EMU values which are comparable to those obtained for the Statz and more basic Curtice model can be obtained using the Curtice-Ettenberg model. This model also produces lower EMU values if only DC data is considered in the EMU formulation.

As expected, the error obtained for a local optimization was better than that obtained for a global optimization. This result is due to the error being calculated only for the saturation region that places less constraints on the model.

Figure 2.44 is a bar graph comparing the newly developed HEMT models using a microwave pseudomorphic HEMT device. The MESFET models from which the HEMT models were derived are also included. Model 8 is derived from MESFET model 5 (see Figure 2.43) by means of the technique described in the previous subsection. Similarly, model 9 is derived from model 3, while model 10 is derived from model 2. Parameter extraction is accomplished in a manner identical to that described in Section 4.3. Both the standard FET models and their corresponding newly developed HEMT models are used in the extraction exercise.

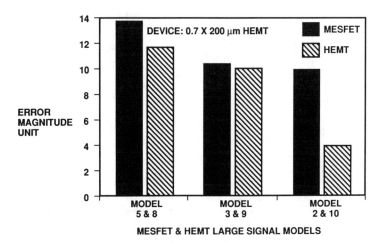

Figure 2.44 The EMU for three MESFET models [16, 18, 20] and the same models with HEMT correction terms added [24]. The models have been applied to a 0.7×200-μm HEMT. Only a global optimization was performed.

The improvement in the model predictions when the three HEMT transconductance degradation terms are added to the advanced Curtice model is clearly evident. Improvements in the predicted dc current-voltage characteristics of Figure 2.45, as well as the predicted microwave transconductance of Figure 2.46 are very significant for this model. Improvements were also made in RF output conductance predictions when the HEMT parameters are added, as shown in Figure 2.47.

The device characteristics of Figures 2.45 through 2.47 are measured for drain-source voltage levels between 0 and 3 V. For the relatively low power HEMTs considered in these studies, this voltage range is adequate. For many devices (especially power devices), however, this characterization range would not be acceptable. For such devices, much larger drain-source voltage levels are required. In general, drain-source bias levels used for device characterization purposes should approach the voltage levels required to achieve gate-drain breakdown.

The error value generated by the parameter extraction program compares very favorably with the best error values found for the MESFET device models. Similar improvements, although of lesser magnitude, are realized for the other two MESFET models modified using the prescribed technique.

The greatest contribution to these improvements comes from the simulation of transconductance. The functional form of the Curtice and advanced Curtice models provides only the possibility of an increasing transconductance with gate-source voltage. The additional degradation term, therefore, has a very large effect.

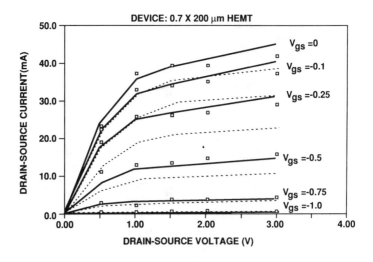

Figure 2.45 Measured and modeled drain-source current as a function of drain-source voltage. Bias points are $V_{gs} = 0$, -0.1, -0.25, -0.5, -0.75, and -1 V. Dashed curves are from model 2, solid curve from model 10; discrete points are measured for a 0.7×200-μm HEMT.

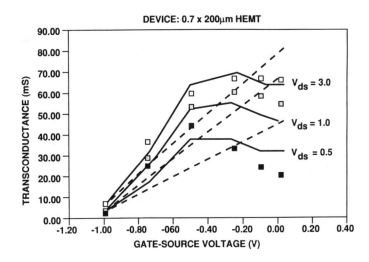

Figure 2.46 Measured and modeled transconductance as a function of gate-source voltage. Bias points are V_{ds} = 0.5, 1, and 3 V. Dashed curves are from model 2, solid curve from model 10; discrete points are measured for a 0.7×200-μm HEMT.

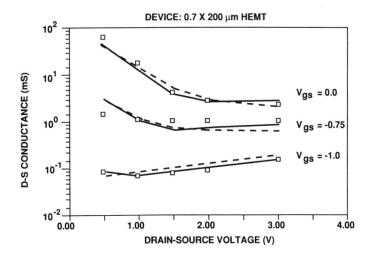

Figure 2.47 Measured and modeled output conductance as a function of drain-source voltage. Bias points are V_{gs} = -1, -0.75, and 0 V. Dashed curves are from model 2, solid curve from model 10; discrete points are measured for a 0.7×200-μm HEMT.

In contrast, only a marginal improvement is gained when the Materka-Kacprzak model is modified. This model has a functional form that allows a peak in transconductance versus gate-source voltage to be predicted without additional terms.

2.3.2 Physically Based Models

Many approaches have been used to model the electrical performance of MESFETs and HEMTs based on their physical characteristics. Some of these involve numerical techniques that are computationally intensive and thus are not useful for most circuit design applications. Analytical models, which are better suited for use in the circuit design process, have also been developed. Two such models are presented in this section: a MESFET model based on that of Lehovec and Zuleeg [26] and a HEMT model originally described by Kola [27]. Although many physically based models have been developed for both MESFETs and HEMTs, few have found their way into commercially available circuit simulation routines or found popular usage among microwave circuit designers. The models presented in this section can be incorporated into large-signal circuit simulators [28] and are somewhat representative of other physically derived models.

2.3.2.1 *Physically Based MESFET Model*

The MESFET model described in this Section is based on the model of Lehovec and Zuleeg [26]. One drawback of their original model is that the computed output conductance is zero in the saturation region. The model can be modified to include a semiempirical gate length modulation, thereby improving the predicted output conductance of the MESFET model in saturation [29]. An additional deficiency of the original model, as with most analytical physical models, is that the computed values of gate-source and gate-drain capacitances are inaccurate. These inaccuracies are due to the simplifying assumptions made in developing analytical expressions for capacitance. For example, the effects of surface states, fringing, and nonabrupt depletion region boundaries are often neglected or inadequately addressed in the development of analytical MESFET capacitance expressions. Models that have addressed these issues typically require knowledge of physical properties that are difficult or impossible to obtain and thus offer no advantages to the circuit designer over empirical models. Rather than present these purely physical models, the use of semiempirical equations for these capacitances has been adopted to improve the accuracy of the model [28].

The parameters used in the equations for this model are listed in Table 2.5. Geometrical, material, and semiempirical parameters are listed separately. The values of the semiempirical parameters are determined through comparison of the modeled to measured performance, but are usually valid for all similar devices fab-

Table 2.5
Physical MESFET Model Parameters

Geometrical parameters:
  a = channel thickness
  L = channel length
  Z = channel width
Material parameters:
  N_d = impurity concentration in channel
  μ_0 = low field mobility in channel
  v_{sat} = saturated drift velocity
Semiempirical parameters:
  Λ = gate-length modulation parameter
  α = subthreshold conduction parameter
Gate-source capacitance:
  C_{GS0} = capacitance at device pinch-off
  C_{max} = capacitance at built-in potential
  m_{GS} = gate-source voltage exponent
  β_{GS} = V_{ds} dependence term in linear region
  λ_{GS1} = V_{gs} dependence term
  λ_{GS2} = V_{ds} dependence term
Gate-drain capacitance:
  C_{GDI} = capacitance factor
  C_{GD0} = capacitance offset
  m_{GD} = gate-drain voltage exponent
  λ_{GD1} = V_{gd} dependence term
  λ_{GD2} = V_{gs} dependence term

ricated with the same process. The number of parameters required is large; however, determination of the device voltage-current behavior and the device capacitance characteristics may be treated independently. As a result, the characterization of the MESFET current, transconductance, and output conductance requires the use of only seven parameters. Fewer parameters are required for each of the capacitances, C_{gs} and C_{gd}.

In addition to the parameters described above, the threshold voltage of the device, V_{TO}, is required as are the applied bias levels, V_{ds} and V_{gs}. The terminal voltages discussed here are the intrinsic voltages (i.e., the voltages that appear internal to the parasitic resistances).

Note that the following equations are derived under the assumption of constant impurity concentration in the channel. This approximation is more accurate for MESFETs fabricated using epitaxial growth techniques than for ion-implanted devices. The epitaxially grown devices demonstrate a fairly constant doping density in the active layer with an abrupt transition to low doping density in the adjacent GaAs substrate or buffer layer. In an ion-implanted device, however, the doping density exhibits a gradual transition from the channel to the substrate. The equations adequately describe ion-implanted devices, however, when effective values of low field mobility μ_0, doping density N_d, and epi-thickness a are used [28].

The drain-source current for a MESFET operating in either the linear or saturated region may be derived from the physical properties and the applied bias voltages following the analysis of Lehovec and Zuleeg [26]. The following expression results:

$$I_{ds} = 4K_T V_{PG}^2 \left[(\xi_D^2 - \xi_S^2) - \tfrac{2}{3} (\xi_D^3 - \xi_S^3) \right] \tag{2.169}$$

where K_T is a proportionality constant, V_{PG} is the channel pinch-off voltage, ξ_D^2 is the normalized gate-drain voltage, and ξ_S^2 is the normalized gate-source voltage. These parameters are defined by the expressions given below.

The proportionality constant is given by

$$K_T = \frac{\mu_0 \epsilon Z}{2aL \left[1 + \nu(\xi_D^2 - \xi_S^2) \right]} \tag{2.170}$$

where μ_0 is the low field channel mobility, ϵ is the permittivity, Z is the gate width, a is the channel thickness, L is the gate length, and ν is the velocity saturation parameter. Values of μ_0 in GaAs are presented in Figure 1.5 as a function of doping density.

The velocity saturation parameter is given by

$$\nu = \frac{\mu_0 V_{PG}}{v_{\text{sat}} L} \tag{2.171}$$

The saturated drift velocity, v_{sat}, is typically 1×10^7 to 2×10^7 cm/s in GaAs. Figure 1.6 illustrates the drift velocity characteristics in several materials as a function of electric field intensity.

The channel pinch-off voltage of the device is a function of the doping density and the channel depth, N_d and a, respectively:

$$V_{PG} = \frac{qN_d a^2}{2\epsilon} \tag{2.172}$$

where the constant, q, is the elementary charge (1.602×10^{-19} C). Note that this quantity is used in the Pucel small-signal and noise model and denoted W_{00} [2].

The normalized drain-source and gate-source voltages may be computed using equations (2.173) and (2.174). The value of V_{DS1} used in the determination of ξ_D depends on whether the device is operating in the linear region or in saturation, and it is defined in equation (2.175):

$$\xi_D = \left(\frac{V_{bi} - V_{gs} + V_{DS1}}{V_{PG}} \right)^{1/2} \tag{2.173}$$

$$\xi_S = \left(\frac{V_{bi} - V_{gs}}{V_{PG}} \right)^{1/2} \tag{2.174}$$

$$V_{DS1} = \begin{cases} V_{ds} & \text{for } V_{ds} \le V_{Dsat} \text{ (linear region)} \\ V_{Dsat} & \text{for } V_{ds} > V_{Dsat} \text{ (saturated region)} \end{cases} \tag{2.175}$$

The saturation voltage, V_{Dsat}, is the value of drain-source potential at which the MESFET enters the saturation region of operation. The procedure for the evaluation of V_{Dsat} is discussed later in this section in equations (2.180) and (2.181).

The built-in gate potential of the MESFET, V_{bi}, is given by the sum of the pinch-off voltage and the threshold voltage, V_{TO}:

$$V_{bi} = V_{PG} + V_{TO} \tag{2.176}$$

The device transconductance and output conductance are obtained by taking the derivative of I_{ds} with respect to V_{gs} and V_{ds}, respectively. The resulting equations depend on the region in which the device is operating. In the linear region ($V_{ds} \le V_{Dsat}$), the transconductance is given by

$$g_m = 4K_T V_{PG}(\xi_D - \xi_S) \tag{2.177}$$

and the output conductance is

$$g_{ds} = \frac{I_{ds}}{K_T} \left(\frac{dK_T}{dV_{ds}} \right) + 4K_T V_{PG}(1 - \xi_D) \tag{2.178}$$

where

$$\frac{dK_T}{dV_{ds}} = \frac{-K_T \nu}{V_{PG}[1 + \nu(\xi_D^2 - \xi_S^2)]} \tag{2.179}$$

Two mechanisms can be responsible for current saturation in the MESFET: channel pinch-off and velocity saturation. To model the behavior of the device, we must determine which of these cases is responsible for saturation at each bias. If sufficient gate bias is applied such that the depletion layer extends through the full depth of the channel, then the current flow restriction is due to channel pinch-off. Current saturation due to velocity saturation occurs when the electric field intensity in the channel is sufficiently high that the electrons moving along the channel achieve their saturated drift velocity, v_{sat}. The velocity-electric field characteristics, for different materials are shown in Figure 1.6, illustrating that the electron velocity in GaAs reaches its maximum at electric field intensities of about 4 kV/cm. Increasing the intensity beyond this value has little effect on the electron velocity. As a result of this relationship between velocity and the electric field, further increases in the drain-source potential have little effect on the current.

In typical microwave devices, that is, in GaAs devices with short gate lengths, velocity saturation is the more common cause of current saturation. In some silicon devices with longer gate lengths, current saturation is caused by pinch-off, except at high current levels. High currents create sufficient electric field intensities for velocity saturation to occur.

To determine which of these mechanisms is responsible for the current saturation, the normalized voltage at which velocity saturation limits the current flow, ξ_{Dvs}, must first be calculated. The value of ξ_{Dvs} is the largest positive real root of the third-order polynomial

$$\xi_{Dvs}^3 + [3(v^{-1} - \xi_S^2)]\xi_{Dvs} + (2\xi_S^3 - 3v^{-1}) = 0 \tag{2.180}$$

Pinch-off of the channel occurs when the normalized gate-drain voltage equals 1; thus, if $\xi_{Dvs} \geq 1$, the current saturation is due to pinch-off. Otherwise velocity saturation is the limiting mechanism. The saturation voltage, V_{Dsat}, is then given by:

$$V_{Dsat} = \begin{cases} \xi_{Dvs}^2 V_{PG} - V_{bi} + V_{gs} & \text{for } \xi_{Dvs} < 1 \\ V_{gs} - V_{TO} & \text{for } \xi_{Dvs} \geq 1 \end{cases} \tag{2.181}$$

The transconductance of a pinched-off device is

$$g_m = \frac{I_{ds}}{K_T}\left(\frac{dK_T}{dV_{gs}}\right) + 4V_{PG}K_T(1 - \xi_S) \tag{2.182}$$

where

$$\frac{dK_T}{dV_{gs}} = \frac{-K_T v}{V_{PG}[1 + v(\xi_D^2 - \xi_S^2)]} \tag{2.183}$$

For the velocity saturation case, the expression for transconductance is more complicated:

$$g_m = \frac{I_{ds}}{K_T}\left(\frac{dK_T}{dV_{gs}}\right) + 4K_T V_{PG}[(2V_{PG}\xi_D\xi_D' + 1) - (2V_{PG}\xi_D^2\xi_D' + \xi_S)] \tag{2.184}$$

with

$$\frac{dK_T}{dV_{gs}} = \frac{-K_T \nu (2V_{PG}\xi_D \xi_D' + 1)}{V_{PG}[1 + \nu(\xi_D^2 - \xi_S^2)]}$$

(2.185)

and with

$$\xi_D' = \frac{d\xi_D}{dV_{gs}} = \frac{-(\xi_D - \xi_S)}{V_{PG}[(\xi_D^2 - \xi_S^2) + \nu^{-1}]}$$

(2.186)

In the calculations above, the computed device current is constant when the device is in saturation, resulting in an output conductance of zero. This is clearly an unrealistic value.

One contributor to the finite output conductance observed in measured data is postulated to be gate length modulation. Complex expressions for gate length modulation effects have been presented by Grebene and Ghandi [30]; however, Hartgring [29] has used a simpler approach in which a correction factor ΔL is subtracted from the channel length:

$$L_{eff} = L - \Delta L$$

(2.187)

The correction term is given by

$$\Delta L = \Lambda a \left(\frac{V_{ds} - V_{Dsat}}{V_{PG}}\right)^{1/2}$$

(2.188)

The semiempirical gate length modulation parameter Λ must be determined experimentally.

When the effects of gate length modulation are incorporated, the output conductance in the saturation region becomes:

$$g_{ds} = \frac{I_{ds}}{L_{eff}}\left(\frac{dL_{eff}}{dV_{ds}}\right)$$

(2.189)

where

$$\frac{dL_{eff}}{dV_{ds}} = \frac{-(\Lambda a)^2}{2V_{PG}\Delta L}$$

(2.190)

Another mechanism that contributes to the output conductance of a MESFET is conduction through the undoped GaAs layer beneath the channel. This behavior may significantly affect the operating characteristics of the device. This effect has been neglected here in order to limit the complexity of the equations.

An additional deficiency of most models is evident at values of V_{gs} near the pinch-off voltage, V_{TO}. This model (as well as most other physically based and empirical large-signal models) predicts lower current near pinch-off than is observed in actual devices. In addition, current continues to flow at $V_{gs} = V_{TO}$ and decreases slowly as the gate bias is made more negative. This behavior is referred to as the *subthreshold effect* and is described in Section 1.4.3.

The device is operating in the subthreshold region when the gate-source voltage is below the threshold onset voltage, V_{TS}, as defined by

$$V_{TS} = V_{TO} + \frac{2\alpha V_{PG}L_{D}}{2a}\left(1 - \frac{\alpha L_{D}}{2a}\right)$$

(2.191)

where α is an empirical parameter and L_D is the intrinsic Debye length. The value of L_D is a function of absolute temperature T, as well as channel doping, and is computed as follows:

$$L_{D} = \left(\frac{\epsilon KT}{q^2 N_d}\right)^{1/2}$$

(2.192)

The device current in the subthreshold region is then obtained by modifying the computed current at $V_{gs} = V_{TS}$ with an exponential factor. The current in the subthreshold region is then given by

$$I_{ds}^{*} = I_{ds(V_{gs}=V_{TS})}\, e^{-K_s(V_{TS}-V_{gs})}$$

(2.193)

where K_s is

$$K_{s} = \frac{g_{m(V_{gs}=V_{TS})}}{I_{ds(V_{gs}=V_{TS})}}$$

(2.194)

Values for the physical and semiempirical parameters extracted for a 1×300-μm MESFET are listed in Table 2.6. The predicted current, transconductance, and output conductance for an extrinsic gate-source bias of -0.8 V and drain-source bias of 3 V are also listed. To demonstrate the effects of parameter variations on the predicted device performance, the device current and conductances have been plotted as a function of gate length, channel thickness, and doping density in Figures 2.48, 2.49, and 2.50, respectively. The dependent and independent parameters in these graphs have been normalized to their nominal values as listed in Table 2.6.

In Figure 2.48, I_{ds}, g_m, and g_{ds} all show an inverse dependence on the gate length; however, the output conductance shows considerably more sensitivity to

Table 2.6
Modeled Parameter Values for Simulated MESFET

a =	0.173 μm
L =	0.653 μm
Z =	300 μm
N_d =	1.29×10^{17} cm^{-3}
μ_0 =	0.402 m^2/V-s
v_{sat} =	2.717 × 10^7 cm/s
Λ =	0.900
α =	0.517
I_{ds} =	31.0 mA
g_m =	45.3 mS
g_{ds} =	1.92 mS

changes in gate length than the others. The transconductance shows a similar trend for variations in channel thickness, as indicated in Figure 2.49. The drain-source current and output conductance, however, are a direct function of the channel thickness. Figure 2.50 illustrates relationships between the performance and doping density of the device that are similar to those observed in Figure 2.49.

Note that the relationships between physical parameters of the device and electrical performance predicted in Figures 2.48 through 2.50 for the Lehovic and

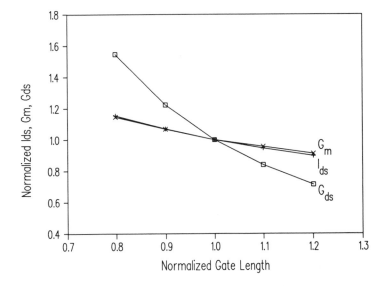

Figure 2.48 Normalized current, transconductance, and output conductance *versus* normalized gate length of a simulated MESFET.

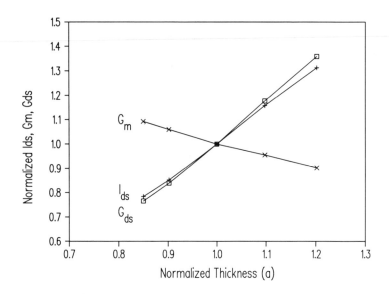

Figure 2.49 Normalized current, transconductance, and output conductance *versus* normalized channel thickness of a simulated MESFET.

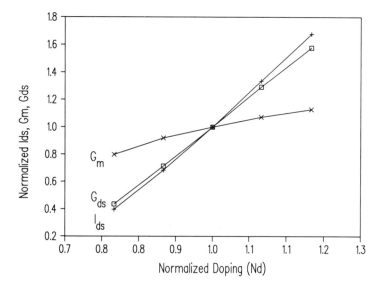

Figure 2.50 Normalized current, transconductance, and output conductance *versus* normalized channel doping of a simulated MESFET.

Zuleeg model are in qualitative agreement with those predicted using the physically based model of Pucel *et al.* (Figures 2.7 through 2.9). Despite this qualitative agreement, significant quantitative differences exist between the two models. These discrepancies indicate that care must be taken in utilizing physically based models to extrapolate results to other devices with drastically different geometries.

The accuracy that can be obtained from the physically based Lehovic and Zuleeg model can be examined by comparing the EMU that results for this model to that obtained by using the previously described empirical models. Application of the parameter extraction algorithm described in Chapter 4 to the same device utilized for the model comparisons of Table 2.5 produces an EMU of 22 for the physically based model—which compares very favorably with both the Curtice model and the Statz model.

In large-signal models, the steady-state current-voltage relationships must be described accurately, and the device capacitance must be represented correctly. Purely physically derived expressions for the device capacitances are typically not accurate enough for use in many important applications. For this reason, the following semiempirical expressions have been developed to describe the capacitance-voltage behavior of the device [28]. The gate-source capacitance is given by:

$$C_{gs} = C_{GS0}Z + \lambda_{GS1} \left(\epsilon ZL/a\right) \left(\frac{V_{gs} - V_{TO}}{V_{bi} - V_{TO}}\right)^{m_{GS}} f_c \qquad (2.195)$$

where

$$f_c = \begin{cases} 1 + \lambda_{GS2} \left(\dfrac{V_{bi} - V_{TO}}{V_{gs} - V_{TO}}\right)^{1/m_{GS}} (V_{ds} - V_{Dsat}) \text{ for } V_{ds} \geq V_{Dsat} \\[2mm] (\lambda_{GS2}V_{Dsat} + \beta_{GS} - 1) \left(\dfrac{V_{ds}}{V_{Dsat}}\right)^2 \\[2mm] + (2 - \lambda_{GS2}V_{Dsat} - 2\beta_{GS}) \left(\dfrac{V_{ds}}{V_{Dsat}}\right) + \beta_{GS} \text{ for } V_{ds} < V_{Dsat} \end{cases} \qquad (2.196)$$

This expression requires that five empirical constants be specified. When capacitance-voltage data are available, these parameters are easily determined. The parameters C_{GS0}, m_{GS}, and λ_{GS1} can be determined from a plot of C_{gs} *versus* V_{gs}, while V_{ds} is held in saturation. The value of C_{GS0} is the value of C_{gs} with a bias of $V_{gs} = V_{TO}$ normalized by the gate width:

$$C_{GS0} = C_{gs(V_{gs}=V_T)}/Z \qquad (2.197)$$

A value C_{max} is also defined from this plot and is given by

$$C_{max} = C_{gs(V_{gs} = V_{bi})}/Z \tag{2.198}$$

The parameter λ_{GS1} can then be computed from

$$\lambda_{GS1} = \frac{C_{max} - C_{GS0}}{(\epsilon L/a)} \tag{2.199}$$

Next, the value of m_{GS} is adjusted to obtain the optimal fit to the data. Alternatively, one can use a parameter extraction algorithm based on optimization, as described in Chapter 4.

The drain-source voltage dependence of the gate-source capacitance is modeled through the function f_c. The parameters λ_{GS2} and β_{GS} determine this dependence. The gate-source capacitance is assumed to be related linearly to V_{ds} when the device is in saturation. The parameter λ_{GS2} is chosen to give the proper dependence in this region. When the device is not in saturation, the capacitance is assumed to have a quadratic dependence on V_{ds}. The value of β_{GS} is chosen to given proper dependence in the nonsaturation region.

The gate-drain capacitance is also described by a semiempirical expression. The equation used for this is

$$C_{gd} = C_{GD}(\epsilon ZL/a) \left(1 - \frac{\lambda_{GD1} V_{gs} - V_{ds}}{\lambda_{GD1} V_{bi}} \right)^{-m_{GD}} (1 - \lambda_{GD2} V_{gs}) + C_{GD0} Z \tag{2.200}$$

In determining the five parameters needed to evaluate this expression, plotting the gate-drain capacitance data as a function of the gate-drain voltage, while keeping the gate-source voltage fixed, is helpful. The parameters C_{GD1}, λ_{GD1}, and m_{GD} determine the dependence of the capacitance on the gate-drain voltage. The parameter λ_{GD2} is then used to obtain the proper gate-source voltage dependence.

2.3.2.2 Physically Based HEMT Model

The HEMT model as described by Kola [27] is based on the physical properties of the device. Approximations to some of the more complex relationships have resulted in an analytical model that may be more useful to microwave circuit design engineers than more complex and time-intensive numerical models. The primary advantage of this model over previous analytical HEMT models is that the Fermi-level variation with electron density in the quantum well is modeled more accurately, resulting in more accurate current-voltage characteristics in the cutoff region of operation. This model was developed for conventional HEMT structures and thus may be less accurate for pseudomorphic HEMTs.

The model is separated into three sections, depending on the device bias conditions and the relationship of the bias to the critical gate voltage V_{GC}. The critical gate voltage is the gate potential at which the electron concentration in the 2-DEG channel reaches the maximum possible sheet density, n_{s0}, for the given heterostructure. The three conditions are:

(1) $V_{gs} < V_{GC}$ (2.201a)

(2) $V_{gs} \geq V_{GC}$ and $V_{gs} - V_{GC} < V_{ds}$ (2.201b)

(3) $V_{gs} \geq V_{GC}$ and $V_{gs} - V_{GC} \geq V_{ds}$ (2.201c)

The electron concentration beneath the gate is illustrated in Figure 2.51 for the three cases listed. In the first case, the electron density, n_s, increases gradually from the source edge of the gate to the drain edge, yet remains less than n_{s0} through the

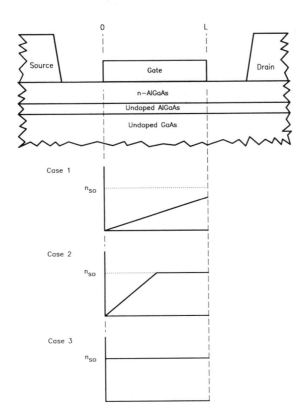

Figure 2.51 Electron sheet density as a function of position along the gate length for the three cases described by equation (2.201).

length of the channel. In case 2, the electron concentration increases until it reaches n_{s0} and then remains constant. If the gate potential is sufficiently high, as in case 3, then the electron density will be at its maximum along the full length of the channel.

The critical gate voltage is determined using the following relation:

$$V_{GC} = \frac{q(d + d_0)n_{s0}}{\epsilon} + V_{off} + E_{F(n_s = n_s0)} \tag{2.202}$$

where q is the elementary charge, d and d_0 are the thickness of the doped layer and undoped AlGaAs layers as shown in Figure 2.52, $E_{F(n_s = n_s0)}$ is the Fermi-level at the maximum electron density, and V_{off} is the voltage at which the 2-DEG layer can no longer exist, and is given by equation (2.206).

The Fermi-level is a nonlinear function of n_s, as illustrated in Figure 2.53. This relationship may be modeled using the following analytical expression [31],

$$E_F = K_1 + K_2(n_s + K_3)^{1/2} \tag{2.203}$$

The parameters K_1, K_2, and K_3 are constants derived by fitting the expression to E_F values obtained using a more accurate and complex function. The values of these constants depend on the type of material used to fabricate the device. For AlGaAs-GaAs structures, the values have been determined empirically to be [31]:

$$K_1 = -0.2083 \tag{2.204a}$$

$$K_2 = 0.3029 \times 10^{-8} \tag{2.204b}$$

$$K_3 = 0.9666 \times 10^{15} \tag{2.204c}$$

The procedure for determining these values for other materials is straightforward and is described in References [27] and [31].

By combining equations (2.202) and (2.203), the critical gate voltage may be written as

$$V_{GC} = \frac{q(d + d_0)n_{s0}}{\epsilon} + V_{off} + K_1 + K_2(n_{s0} + K_3)^{1/2} \tag{2.205}$$

The voltage at which the 2-DEG layer is "turned off," V_{off}, is given by

$$V_{off} = V_{bi} - V_{P2} - \Delta E_c \tag{2.206}$$

where V_{bi} is the Schottky barrier height on the wide-band-gap material and ΔE_c is the conduction band discontinuity between the wide- and narrow-band-gap materials. Both V_{bi} and ΔE_c are constants. For a conventional AlGaAs-GaAs HEMT,

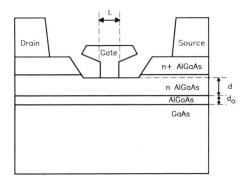

Figure 2.52 Cross section of a HEMT illustrating important dimensions.

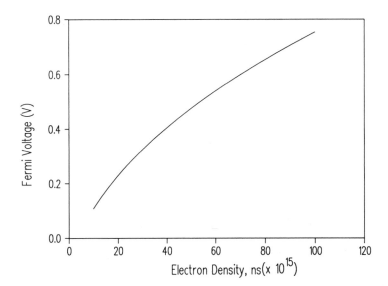

Figure 2.53 Fermi voltage as a function of electron density for the AlGaAs-GaAs HEMT structure.

these constants have typical values of 1 V and 0.24 eV, respectively. The conduction band discontinuity, however, is a function of the proportion of aluminum to gallium in the AlGaAs layer. If this ratio is known, more accurate values of ΔE_c may be obtained [32]. The AlGaAs pinch-off potential, V_{P2}, is calculated by

$$V_{P2} = \frac{qN_d d^2}{2\epsilon} \tag{2.207}$$

where N_d is the donor density in the wide-band-gap material.

To summarize, the necessary equations to determine the critical gate voltage are (2.205) through (2.206). The following constants represent known quantities: q, ϵ, V_{bi}, ΔE_c, K_1, K_2, and K_3. Four geometric dimensions or material parameters are required: d, d_0, n_{s0}, and N_d.

An important parameter that must be determined prior to calculation of device current for cases 1 and 2 is the drain-source saturation voltage, V_{Dsat}. This value is obtained by solving the fourth-order polynomial:

$$V_{DSAT}^4 + 4(\epsilon L + C/9)V_{DSAT}^3 + 4[\epsilon^2 L^2 + (D + 2C\epsilon L)/3 - P]V_{DSAT}^2$$
$$+ 4[C\epsilon^2 L^2 + \epsilon L(2D/3 - 2P)]V_{DSAT}$$
$$+ 4(P^2 - 4D^3/9C^2 - 4D^2\epsilon L/3C - D\epsilon^2 L^2) = 0 \qquad (2.208)$$

The value of P is dependent on the bias conditions as follows:

$$P = \begin{cases} P_0 & \text{for } V_{gs} < V_{GC} \text{ (case 1)} \\ P_0 + K_I & \text{for } V_{gs} \geq V_{GC} \text{ and } V_{gs} - V_{GC} < V_{ds} \text{ (case 2)} \end{cases} \qquad (2.209)$$

where P_0 and K_I are given by:

$$P_0 = (V_{GI} + C/2)\epsilon L + 2D^{3/2}/3C \qquad (2.210)$$

$$K_I = \frac{(V_{gs} - V_{GC})^2}{2} + \frac{2}{3C}\{[D - C(V_{gs} - V_{GC})]^{3/2} - D^{3/2}\}$$
$$+ \left[\frac{C^2}{4} + C(V_{GC} - V_{off} - K_1) + K_2^2 K_3\right]^{1/2}(V_{gs} - V_{GC}) \qquad (2.211)$$

The other parameters required are given by the following equations:

$$V_{GI} = V_{gs} - V_{off} - K_1 \qquad (2.212)$$

$$A = \frac{\mu_0 Z \epsilon}{L(d + d_0)} \qquad (2.213)$$

$$B = 1 + \frac{V_{ds}}{\epsilon L} \qquad (2.214)$$

$$C = \frac{\epsilon K_2^2}{q(d + d_0)} \qquad (2.215)$$

$$D = \frac{C^2}{4} + CV_{GI} + K_2^2 K_3 \qquad (2.216)$$

where ϵ is the dielectric constant of the substrate, μ_0 is the low field electron mobility in the two-dimensional electron gas, Z is the gate width, and L is the gate length.

The current, transconductance, and output conductance of the HEMT are determined by using the following equations:

$$I_{ds} = I_{DS0} (1 + \Lambda V_{ds}) \tag{2.217}$$

$$g_m = G_{m0} (1 + \Lambda V_{ds}) \tag{2.218}$$

$$g_{ds} = G_{DS0} (1 + \Lambda V_{ds}) + \Lambda I_{DS0} \tag{2.219}$$

These expressions include the effects of gate length modulation, which increases the drain-source current and provides a means for modeling the finite output conductance that occurs in actual devices. The output conductance parameter Λ is an empirical value determined by fitting the model to the measured data. The parameters I_{DS0}, G_{m0}, and G_{DS0} describe the device behavior without the gate length modulation effects. The expressions used to define these parameters depend on the gate-source bias conditions as described by equation (2.201), as well as on the region of operation (linear or saturation) of the device.

When the device is biased such that V_{gs} is less than V_{GSC} (case 1), and is in the linear region of operation, the expression for I_{DS0} is

$$I_{DS0} = \frac{A}{B} \left(V_{G1} V_{ds} + \frac{C V_{ds}}{2} - \frac{V_{ds}^2}{2} + \frac{2\{(D - C V_{ds})^{3/2} - D^{3/2}\}}{3C} \right) \tag{2.220}$$

In the saturation region ($V_{ds} > V_{Dsat}$), I_{DS0} is assumed to be independent of drain-source voltage, and it remains constant at the value of I_{DS0} calculated at $V_{ds} = V_{DSAT}$.

$$I_{DS0(V_{ds} \geq V_{Dsat})} = I_{DS0(V_{ds} = V_{dsat})} \tag{2.221}$$

The transconductance term G_{m0} for the linear and saturation regions is given by

$$G_{m0} = \begin{cases} \dfrac{A}{B} \{V_{ds} + [(D - C V_{ds})^{1/2} - D^{1/2}]\} & \text{for } V_{ds} < V_{Dsat} \\[2mm] \dfrac{A}{B} \{V_{Dsat} + [(D - C V_{DSAT})^{1/2} - D^{1/2}]\} & \text{for } V_{ds} \geq V_{Dsat} \end{cases} \tag{2.222}$$

To determine G_{DS0} for case 1, the following two relations are used for the linear and saturation regions, respectively:

$$G_{DS0} = \frac{A \epsilon L}{(\epsilon L + V_{ds})^2} \left\{ \epsilon L \left(V_{G1} + \frac{C}{2} \right) - V_{ds} \left(\epsilon L + \frac{V_{ds}}{2} \right) \right.$$

$$\left. - (D - C V_{ds})^{1/2} \times (\epsilon L + V_{ds}) - \frac{2}{3C} [(D - C V_{ds})^{3/2} - D^{3/2}] \right\} \tag{2.223a}$$

$$G_{DS0} = 0 \text{ for } V_{ds} \ge V_{DSAT} \tag{2.223b}$$

When the device is operating under the conditions of case 2, $V_{gs} - V_{GC} < V_{ds}$, I_{DS0} is related to that of case 1 through the following relationship:

$$I_{DS0} = I'_{DS0} - \frac{AK_I}{B} \tag{2.224}$$

where I'_{DS0} is the current for case 1 as given by equations (2.220) and (2.221).

Equation (2.225) is used to determine G_{m0} in the linear region. To determine the transconductance term when $V_{ds} > V_{Dsat}$, V_{ds} is replaced by the value of V_{Dsat}:

$$G_{m0} = \frac{A}{B}\{V_{ds} + [(D - CV_{ds})^{1/2} - D^{1/2}]\} - \frac{A}{B}\left\{(V_{gs} - V_{GC}) - D^{1/2}\right.$$
$$\left. - \left[\frac{C}{4} + C(V_{GC} - V_{off} - K_1) + K_2^2K_3\right]^{1/2}\right\} \tag{2.225}$$

The output conductance is given by

$$G_{DS0} = G'_{DS0} + \frac{A\epsilon L}{(\epsilon L + V_{ds})^2} K_I \tag{2.226}$$

where G'_{DS0} is calculated using equation (2.223).

For case 3, the electron density remains constant at the value of n_{s0} through the length of the channel. The drain-source current for this situation is given by

$$I_{DS0} = \frac{\mu_0 Zqn_{s0}V_{ds}}{LB} \tag{2.227}$$

The current density is unaffected by V_{gs} in this case, thus the transconductance G_{m0} is zero. The output conductance is given by

$$G_{DS0} = \frac{\mu_0 Zqn_{s0}}{LB^2} \tag{2.228}$$

Parallel conduction through the wide-band-gap material (AlGaAs in a conventional HEMT) occurs when the material is not fully depleted. This behavior is referred to as conduction via the parasitic MESFET. In an enhancement mode device, this occurs when the potential on the gate exceeds the critical gate voltage. In a depletion mode device in which the AlGaAs layer is relatively thick, the depletion layers formed by the Schottky barrier and the heterojunction interface do not overlap for zero gate bias. The parasitic MESFET effect may be modeled by utilizing a standard MESFET model in parallel with the HEMT model [27], although such a technique is a simplification of the actual physical processes taking place in the device.

Capacitance expressions for HEMTs have been developed [33] that relate the gate-source and gate-drain capacitances to the physical characteristics of the device and the applied bias conditions. These capacitances are assumed to be independent of bias for values of V_{ds} greater than the saturation voltage. The source-drain capacitance is assumed to be constant for all bias levels. Such assumptions are approximately consistent with the observed HEMT characteristics (see Chapter 1).

The resulting gate-source and gate-drain capacitance expressions are very similar to those presented as empirical HEMT capacitance expressions:

$$
C_{gs} = \begin{cases} \dfrac{2C_G}{3}\left[1 - \dfrac{(V_{Dsat} - V_{ds})^2}{(2V_{Dsat} - V_{ds})^2}\right] + C_{GS0} & \text{for } V_{ds} < V_{Dsat} \\[4mm] \dfrac{2C_G}{3} + C_{GS0} & \text{for } V_{ds} \geq V_{Dsat} \end{cases} \tag{2.229}
$$

The gate-drain capacitance is given by

$$
C_{gd} = \begin{cases} \dfrac{2C_G}{3}\left[1 - \dfrac{V_{dsat}^2}{(2V_{Dsat} - V_{ds})^2}\right] + C_{GD0} & \text{for } V_{ds} < V_{Dsat} \\[4mm] C_{GD0} & \text{for } V_{ds} \geq V_{Dsat} \end{cases} \tag{2.230}
$$

The constants C_{GS0} and C_{GD0} are most easily determined experimentally. The parameter C_G is the total gate depletion capacitance and is obtained through the following relations:

$$
C_G = \left(\frac{1}{C_1} + \frac{1}{C_2} + \frac{1}{C_3}\right)^{-1} \tag{2.231}
$$

where

$$
C_1 = \epsilon/d, \tag{2.232}
$$

$$
C_2 = q^2 D_0[f(E_0) + f(E_1)], \text{ and} \tag{2.233}
$$

$$
C_3 = \frac{3q^2 n_s\, [f(E_0) + f(E_1)]}{2[E_0 f(E_0) + E_1 f(E_1)]} \tag{2.234}
$$

Here D_0 is the density of states of the electron gas and has been determined experimentally to have a value of 2.025×10^{36} J/m². The parameters E_0 and E_1 are the energy levels of the first two subbands and are related to the electron density as follows:

$$E_0 = 2.5 \times 10^{-12} n_s^{2/3} \tag{2.235}$$

$$E_1 = 3.2 \times 10^{-12} n_s^{2/3} \tag{2.236}$$

The function $f(E_n)$ is given by

$$f(E_n) = \left[1 + e^{(E_n - E_F/kT)} \right]^{-1} \tag{2.237}$$

where E_F is the Fermi voltage found in equation (2.203).

REFERENCES

[1] J.M. Golio, "Compound Semiconductor for Microwave MESFET Applications," MS Thesis, North Carolina State University, 1980.

[2] R.A. Pucel, H.A. Haus, and H. Statz, "Signal and Noise Properties of Gallium Arsenide Microwave Field-Effect Transistors," *Advances in Electronics and Electron Physics,* Vol. 38, New York: Academic Press, pp. 195–265.

[3] J.M. Golio, "Ion-Implanted MESFETs," Ph.D. Dissertation, North Carolina State University, 1983.

[4] J.M. Golio and R.J. Trew, "Profile Studies of Ion-Implanted MESFETs," *IEEE Trans. Microwave Theory Tech.,* Vol. MTT-31, December 1983, pp. 1066–1071.

[5] J.M. Golio, M. Miller, G.N. Maracas, and D.A. Johnson, "Frequency-Dependent Electrical Characteristics of GaAs MESFETs," *IEEE Trans. Electron Devices,* Vol. ED-37, May 1990, pp. 1217–1227.

[6] N.J. Kuhn, "A Survey of Transistor Noise Characterization," presented at HP RF and Microwave Symposium, Phoenix, Arizona, September 1986.

[7] G. Gonzalez, *Microwave Transistor Amplifiers, Analysis and Design,* Englewood Cliffs, NJ: Prentice-Hall, Inc., 1984, pp. 139–154.

[8] R. Pengelly, *Microwave Field-Effect Transistors—Theory, Design and Applications,* Chichester, UK: Research Studies Press, 1982, pp. 28–39.

[9] A.F. Podell, "A Functional GaAs FET Noise Model," *IEEE Trans. Electron Devices,* Vol. ED-28, May 1981, pp. 511–517.

[10] M.S. Gupta, O. Pitzalis, Jr., S.E. Rosenbaum, and P.T. Greiling, "Microwave Noise Characterization of GaAs MESFET's: Evaluation by On-Wafer Low-Frequency Output Noise Current Measurement," *IEEE Trans. Microwave Theory Tech.,* Vol. MTT-35, December 1987, pp. 1208–1218.

[11] H. Fukui, "Design of Microwave GaAs MESFET's for Broad-Band Low-Noise Amplifiers," *IEEE Trans. Microwave Theory Tech.,* Vol. MTT-27, July 1979, pp. 643–650.

[12] H. Fukui, "Optimal Noise Figure of Microwave GaAs MESFET's" *IEEE Trans. Electron Devices,* Vol. ED-26, July 1979, pp. 1032–1037.

[13] A. Cappy, A. Vanoverschelde, M. Schortgen, C. Versnaeyen, and G. Salmer, "Noise Modeling in Submicrometer-Gate Two-Dimensional Electron-Gas Field-Effect Transistors," *IEEE Trans. Electron Devices,* Vol. ED-32, December 1985, pp. 2787–2795.

[14] H.S. Statz, H.A. Haus, and R.A. Pucel, "Noise Characteristics of Gallium Arsenide Field-Effect Transistors," *IEEE Trans. Electron Devices,* Vol. ED-21, September 1974, pp. 549–562.

[15] T.M. Brookes, "The Noise Properties of High Electron Mobility Transistors," *IEEE Trans. Electron Devices,* Vol. ED-33, January 1986, pp. 52–57.

[16] W.R. Curtice, "A MESFET Model for Use in the Design of GaAs Integrated Circuits," *IEEE Trans. Microwave Theory Tech.,* Vol. MTT-28, 1980, pp. 448–456.

[17] H. Statz, P. Newman, I. Smith, R. Pucel, and H. Haus, "GaAs FET Device and Circuit Simulation in SPICE," *IEEE Trans. Electron Devices,* Vol. ED-34, 1987, pp. 160–169.

[18] T. Kacprzak and A. Materka, "Compact dc Model of GaAs FET's for Large-Signal Computer Calculation," *IEEE J. Solid-State Circuits,* Vol. SC-18, April 1983, pp. 211–213.

[19] A. McCamant, G. McCormack, and D. Smith, "An Improved GaAs MESFET Model for Spice," *IEEE Microwave Theory Tech.,* Vol. MTT-38, June 1990, pp. 822–824.

[20] *HSPICE User's Manual,* Version H8907, Meta-Software, Campbell, CA, 1989.

[21] W.R. Curtice and M. Ettenberg, "A Nonlinear GaAs FET Model for Use in the Design of Output Circuits for Power Amplifiers," *IEEE Trans. Microwave Theory Tech.,* Vol. MTT-33, 1985, pp. 1383–1394.

[22] R. Van Tuyl and C. Liechti, "Gallium Arsenide Digital Integrated Circuits," Technical Report AFL-TR-74-40, Air Force Avionics Lab., March 1974.

[23] T. Taki, "Approximation of Junction Field-Effect Transistor Characteristics by a Hyperbolic Function," *IEEE J. Solid-State Circuits,* Vol. SC-13, October 1978, pp. 724–726.

[24] M. Miller, M. Golio, B. Beckwith, E. Arnold, D. Halchin, S. Ageno, and S. Dorn, "Choosing an Optimum Large Signal Model for GaAs MESFETs and HEMTs," *IEEE Microwave Theory Tech. Symp. Digest,* 1990, pp. 1279–1282.

[25] J. Meyer, "MOS Models and Circuit Simulation," *RCA Rev.,* Vol. 32, March 1971, pp. 42–63.

[26] K. Lehovic and R. Zuleeg, "Voltage-Current Characteristics of GaAs J-FET's in the Hot Electron Range," *Solid-State Electron.,* Vol. 13, 1970, pp. 1415–1426.

[27] S. Kola, "An Analytic Modelling of HEMTs," MS Thesis, Arizona State University, 1987.

[28] J.M. Golio, J.R. Hauser, and P.A. Blakey, "A Large-signal GaAs MESFET Model Implemented on SPICE," *IEEE Circuits and Devices Mag.,* September 1985, pp. 21–30.

[29] C.D. Hartgring, "Silicon MESFET's," Ph.D. Dissertation, University of California, Berkeley, 1981.

[30] A.B. Grebene and S.K. Ghandi, "General Theory for Pinched Operation of the Junction-Gate FET," *Solid-State Electron.,* Vol. 12, No. 7, July 1969, pp. 573–588.

[31] S. Kola, J.M. Golio, and G.N. Maracas, "An Analytical Expression for Fermi-Level Versus Sheet Carrier Concentration for HEMT Modeling," *IEEE Electron Devices Lett.,* Vol. EDL-9, No. 3, March 1988, pp. 136–138.

[32] P.H. Ladbrooke, "MMIC Design: GaAs FETs and HEMTs," Norwood, MA: Artech House, 1989, pp. 190–192.

[33] G. George and R. Hauser, "An Analytic Model for MODFET Capacitance Voltage Characteristics," *IEEE Trans. Electron Devices,* Vol. ED-37, No. 5, May 1990, pp. 1193–1198.

Chapter 3
CHARACTERIZATION

An important relationship exists among device models (the subject of Chapter 2), parameter extraction techniques (the subject of Chapter 4), and device measurements (the subject of this chapter). The accuracy of all models is determined not only by the form and features of the model, but also by the validity of the parameters used within the model. Reliable determination of these parameters for any finished device requires some device characterization. The accuracy and relevancy of that characterization contributes directly to the reliability of the parameters and therefore to the accuracy of the model. This is true even for pure physically based models. A physically based model requires information regarding device dimensions and material parameters. Gate length and doping density are examples of parameters typically required of a physically based model. In a finished device, however, the exact gate length and doping density are not known without some measurement or characterization.

For empirical models, the required characterization is almost always a measurement of electrical performance. Small-signal *S*-parameters, dc current-voltage measurements, load-pull measurements, measurements of output harmonic content as a function of input power, and noise figure measurements are typical of device characterization used with empirical modeling. Some of these measurements are fairly simple and straightforward to obtain using standard test equipment, while others can become tedious and involve expensive equipment configurations.

The measurements required to obtain accurate information about the physical parameters of a device are often more difficult to perform than measurements of electrical parameters. The dimensions of contacts and spacings on the surface of the device, the gate length, for example, can be measured using a scanning electron microscope. To perform this measurement, however, layers of material used to passivate the device must first be removed. The measurement is therefore destructive. Doping density profiles of material can be approximately determined using C-V measurement techniques. Unfortunately, the C-V techniques require simultaneous fabrication of a specialized test structure on the same wafer with the device. Such a structure is seldom available to the circuit designer.

This chapter discusses only electrical characterization techniques. Measurements of dc characteristics and small-signal S-parameters are among the most routine and important measurements used to extract model parameters, and they are discussed in the first two sections of this chapter. Other characterization techniques—used primarily for large-signal applications—presented in this chapter include pulsed current-voltage characterization, load-pull, and measured harmonic content as a function of input power. Finally, the measurements required to utilize noise figure models are examined. These include both simple-to-make noise spectral density measurements as well as fully automated noise figure measurements. A method for using the electrical characterization data to extract physically based model parameters is presented in Chapter 4.

3.1 DIRECT CURRENT MEASUREMENTS

3.1.1 Conventional Current-Voltage Measurements

The primary advantage of utilizing dc data for model parameter extraction purposes is that dc measurements are easy to perform. In addition, device characteristics such as output conductance and transconductance can be estimated from the drain-source current *versus* the drain-source voltage and drain-source current *versus* gate-source voltage, respectively. These I-V measurements are easily automated using computer-controlled test equipment. An automated test configuration of this type is illustrated in Figure 3.1. The measurement is very simple to make and requires only commonly available, low-cost instruments. The power supplies and meters must function over the range of voltages and currents determined by the device characteristics. For typical microwave low noise or medium power devices, these requirements seldom exceed approximately 15 V and 300 mA. For power devices, higher voltages and currents are often required.

One difficulty that sometimes arises when attempting to make dc measurements is device instability. Microwave devices have a tendency to become unstable when connected to a measurement system of conflicting impedances. Many high gain devices exhibit unstable characteristics even when carefully embedded in a matched 50-Ω measurement system. These instabilities affect the current-voltage characteristics of the MESFET or HEMT. For this reason, care must be taken to ensure that the device is stable during the I-V measurements. Devices with an appropriate on-wafer terminal pattern (see Figure 3.2) are measured using a microwave wafer probing station similar to that pictured in Figures 3.3(a) and 3.3(b). Microwave probe stations use coplanar transmission line probes, which bring 50-Ω lines directly to the device terminals on a wafer. Use of a fixed impedance system

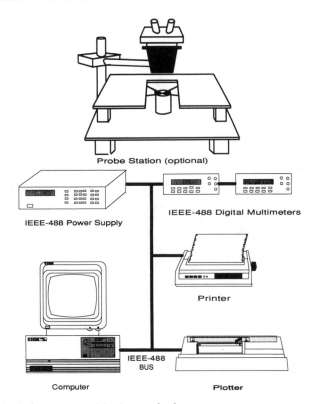

Figure 3.1 Diagram of an automated dc characterization system.

Figure 3.2 Photograph of an on-wafer transistor that can be probed. The large pads at each corner of the pattern are grounded using via holes. Coplanar probes make contact with these ground pads as well as with the signal lines.

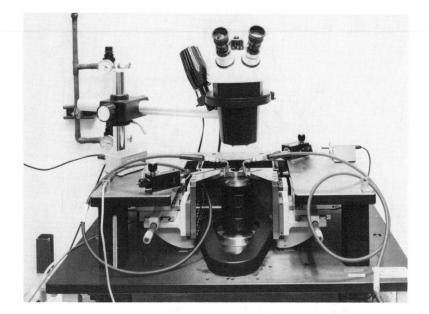

(b)

Figure 3.3 Photograph of a microwave probing station: (a) probe station and microscope, (b) closer view of coplanar probes.

such as this ensures excellent signal grounds and helps eliminate many unwanted device oscillations. An added advantage of microwave probe stations is that they can be used to measure both dc and RF characteristics simultaneously, as discussed later in this section.

Chip and discrete devices can be measured using a microstrip test fixture such as that pictured in Figure 3.4. These test fixtures typically have coaxial connectors and a microstrip-to-device connection. An impedance discontinuity occurs at both the coaxial-to-microstrip launch point and the microstrip-to-device connection. In addition, the signal is attenuated in both the coaxial and microstrip portions of the test fixture. These effects must be removed through calibration or de-embedding before accurate RF measurements can be made. These fixtures can also be used to help prevent instability in the device when measuring its dc characteristics.

Figure 3.4 Photograph of a microwave microstrip test fixture used to characterize discrete devices.

Figures 3.5(a) and 3.5(b) present measured drain-source current as a function of both applied drain-source and gate-source bias levels for a 0.5×300-μm gate length MESFET. These measurements are made using an automated test configuration similar to that of Figure 3.1. By definition, the output conductance of the device is given by the derivative of the drain-source current with respect to drain-source voltage. Likewise, transconductance can be expressed as the derivative of the drain-source current with respect to gate-source voltage. Because of the fre-

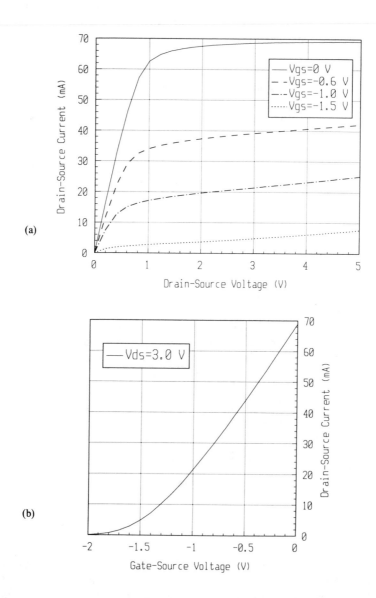

Figure 3.5 Measured results of dc current-voltage characteristics of a microwave MESFET using the measurement configuration of Figure 3.1: (a) drain-source current *versus* drain-source voltage, (b) drain-source current *versus* gate-source voltage.

quency dependence of these characteristics, the curves of Figures 3.5(a) and 3.5(b) alone cannot provide adequate characterization for use in parameter extraction. Note that although the output conductance varies significantly as frequency is increased from dc to approximately 1 MHz, the actual measured currents and the transconductance vary with frequency by a much smaller amount. The data presented in Figure 3.5, therefore, may be useful in obtaining first-order estimates of RF device performance characteristics.

The dc output resistance is determined by incrementing the drain-source bias voltage in 0.1-V steps and then measuring the difference in current through a resistor R (see Figure 3.6). This process must be done using a delay of several seconds between each bias point measurement. This delay is essential to obtain accurate dc data because device characteristics can drift for times on the order of several milliseconds. The dc transconductance can be determined in a similar manner.

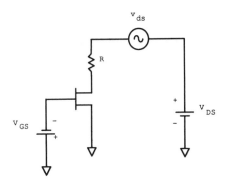

Figure 3.6 Measurement configuration used to determine the dc output resistance of a MESFET or HEMT.

3.1.2 Forward-Bias Gate Measurements

The parasitic resistance values R_s, R_d, and R_g, found in MESFET equivalent circuits, can be estimated from the analysis of three separate dc current-voltage measurements [1]. The required measurements utilize properties of the forward-biased Schottky gate. Although the technique does not predict accurate parasitic resistance values for the HEMT, it does work well for MESFETs. The reason for the failure of the method when applied to HEMTs is discussed later in this section.

The forward gate current as a function of the applied forward gate potential is presented in Figure 3.7. At low voltages, the gate operates as an ideal forward-biased diode, with a gate current given by:

$$I_g = I_s[e^{(qV/nkT)} - 1], \tag{3.1}$$

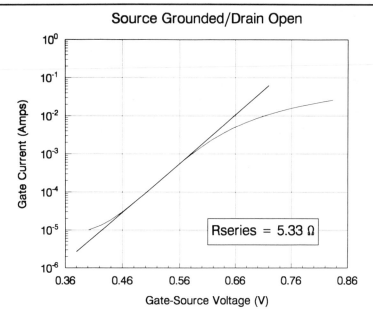

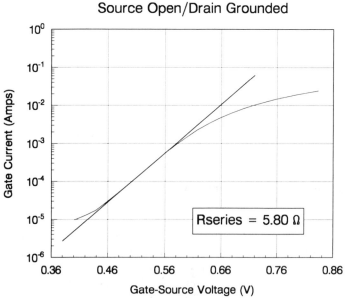

Figure 3.7 Measured results of forward-bias gate measurements illustrated in Figure 3.8. Results are obtained for a microwave MESFET: (a) drain terminal open and source terminal grounded, (a) source terminal open and drain terminal grounded, (c) both drain and source terminals grounded.

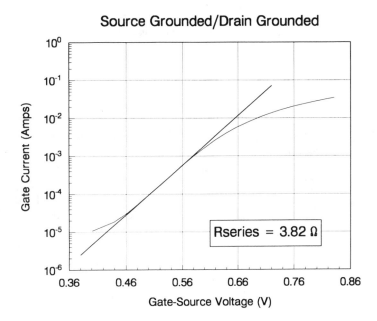

Source Grounded/Drain Grounded

Rseries = 3.82 Ω

Figure 3.7 continued

or, in terms of the applied forward potential,

$$V = (nkT/q) \ln[I_g/I_s + 1] \qquad (3.2)$$

where I_s is the reverse saturation current, I_g is the gate current, q is the electron charge, V is the applied forward potential of an ideal diode, n is the ideality factor, k is Boltzmann's constant, and T is the device temperature (in degrees Kelvin).

In the ideal diode, the $I_g - V_g$ relationship maintains a constant slope on a semilog plot, as indicated by the straight line in Figure 3.7. At higher current levels, however, the voltage drop across the parasitic resistances becomes significant with respect to the diode voltage drop, and the curve deviates from this line. The potential across the gate V_g, including any parasitic resistances in series with it, is written as:

$$V_g = V + I_g R_{ser} \qquad (3.3)$$

where R_{ser} is the total resistance in series with the gate diode. This variation from the ideal diode performance is easily observed in Figure 3.7, and it is this deviation that provides the necessary information to allow extraction of the parasitic resistance values. The total series resistance is obtained as follows:

$$R_{ser} = \frac{(V_{g0} - V_{i0})}{I_{g0}}$$

(3.4)

where V_{g0} is the measured gate voltage at current I_{g0} and V_{i0} is the voltage drop across the ideal gate diode at current I_{g0} and is obtained by extrapolating the linear portion of the measured data to the desired current.

Three simple measurements are used to obtain R_{ser} as illustrated in Figures 3.8(a) through 3.8(c). These measurements involve the forward conduction properties of the gate electrode of the MESFET. A simpler form of the technique has been used in diode characterization and is easily applied to the MESFET problem. One measurement, illustrated in Figure 3.8(a), is made by leaving the circuits on the drain terminal open. A second measurement [Figure 3.8(b)] is made with the source circuits left open. The final measurement [Figure 3.8(c)] is made with the drain and source grounded. Each measurement configuration provides a different R_{ser}, which can then be used to obtain the individual parasitic resistance values using the technique described in Section 4.1.

Figures 3.7(a) through 3.8(c) are examples of forward-bias data for a $0.5 \times$ 300-μm MESFET. Figures 3.7(a) and 3.7(b) are graphs of the gate current *versus* the applied positive gate voltage with the drain circuits open and the source grounded, and with the drain grounded and the source circuits open, respectively. Figure 3.7(c) is a graph of the gate current *versus* the applied positive gate voltage with the source and drain grounded.

Any of these three forward-bias measurements can also be used to extract the parameters required to describe the gate-drain and gate-source diodes of the large-signal equivalent circuit of Figure 2.33. Although these diode parameters apply to both MESFET and HEMT models, parasitic resistances determined using forward-bias gate measurements will not produce accurate values for the HEMT. Details concerning the extraction of all of these parameters from raw data are provided in Chapter 4.

3.2 RADIO-FREQUENCY MEASUREMENTS

The dc measurements alone are not sufficient to characterize microwave device completely. Small-signal model element values must be determined from RF impedance measurements. Frequency-dependent characteristics such as the shift in output conductance and transconductance can only be accounted for by making

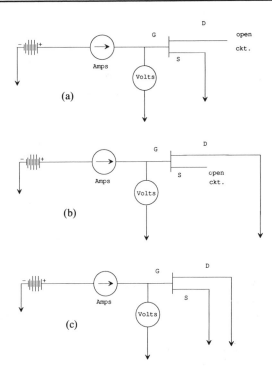

Figure 3.8 Forward-bias gate measurement data. The measurements provide estimates of diode parameters for both the MESFET and HEMT and parasitic resistance values for the MESFET: (a)
drain terminal open and source terminal grounded; (b) source terminal open and drain terminal grounded; (c) both drain and source terminals grounded.

measurements covering frequencies from just above dc to RF. Typical RF measurements include S-parameter measurements and harmonic content measurements. Pulsed measurements can also be used to derive microwave frequency
device characteristics as well as avalanche properties of the gate. Data resulting
from these measurements can then be used to derive small- and large-signal device
models.

3.2.1 Microwave S-Parameter Measurements

The most common RF measurements utilized for the characterization of
MESFETs and HEMTs for parameter extraction purposes are microwave S-param-

eter measurements. Conventional S-parameter measurements characterize the small-signal performance of a device at one bias point. To obtain large-signal information, S-parameter characterization must be performed on the device over a large range of operating bias levels [2]. Typically, the S-parameters corresponding to each particular bias level are used to determine a unique set of values for small-signal equivalent circuit elements. This information is then used to determine the dependence of each equivalent circuit element on bias conditions. The process of extracting the circuit element values from measured S-parameters is detailed in Section 4.2.

When S-parameter data are to be used to extract large-signal modeling information, the measurements need to be performed over an appropriate bias range. Typically, these measurements should be performed at 15 to 25 bias points. The gate-source voltage should be varied from nearly pinch-off to slightly forward biased. Drain-source voltage settings should include some low bias values in the linear region of the I-V curves, some bias levels near the knee of the I_{ds}-V_{ds} curve, and several bias levels with the device in saturation. The maximum drain-source voltage to use is determined by the breakdown voltage of the device. Most of the devices used to compile the data presented in this book have breakdown voltages of approximately 7 V or less. For these low and medium power devices, drain-source bias levels are chosen between 0.5 and 5.0 V. For larger power devices, drain-source bias levels much higher than this should be selected.

Automated measurement equipment that performs S-parameter measurements at multiple bias settings is readily available in most laboratories. Figure 3.9 depicts a typical automated test configuration. The computer controls the network analyzer, multiple output power supply, and multimeters. A configuration of this type allows both the S-parameter data and dc data to be acquired simultaneously. The power supplies provide desired bias settings for the device and these settings are monitored using the multimeters. Consideration of the resistive losses inherent to the measurement equipment must be made when interpreting the bias readings. The network analyzer is clearly the heart of this simple configuration. Modern network analyzers have made the process of collecting S-parameter data over frequency a relatively trivial task. Special device information (i.e., manufacturer, manufacturing process, gate area, *et cetera*) as well as the measured bias voltages and currents can be stored in a header for each S-parameter file for later use in the parameter extraction and modeling processes.

Measuring device S-parameters in a two-port configuration can be an extremely accurate process, even beyond 60 GHz, due to the sophistication of the network analyzers in use today. The accuracy, however, depends largely on the ability to calibrate the system and de-embed the device test fixture.

Two techniques commonly used to calibrate network analyzers and wafer probing stations are the short-open-load through (SOLT) and the through-reflect-line (TRL) methods. A typical two-port calibration uses a short, open, and load at

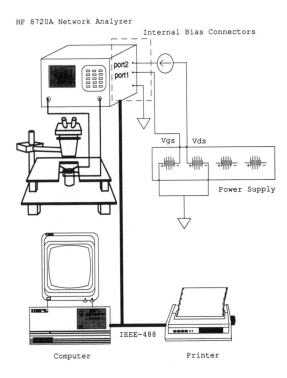

Figure 3.9 Automated *S*-parameter test equipment configuration; the figure illustrates a system used with a microwave probe station.

each port as impedance standards to measure reflection errors, a through standard to measure frequency response tracking and load match, and an open signal path to measure isolation between ports. This type of calibration is typical for a coaxial transmission medium. When the transmission medium is not coaxial, such as in a microstrip, the calibration standards of the SOLT method are extremely difficult to fabricate, which hampers the ability to make an accurate calibration. For this reason, the TRL method should be considered as an alternative calibration method [3]. The TRL calibration consists of a through of zero length (ideally zero length, but it may also be a known, nonzero length) and zero loss. The purpose of this standard is to establish a reference plane. The reflect standard is used to contribute the same unknown high reflection at both test ports. The line is used to determine the reference impedance of the measurement. The line has a length different from the through and is assumed to be reflectionless.

Table 3.1 demonstrates the relative accuracy of the error correction for the SOLT and TRL calibrations [4]. The SOLT calibration is realized with a fixed load, a sliding load, and an offset load. The SOLT calibration with fixed load is one of

Table 3.1
Accuracy of Error Correction (7-mm Connectors at 18 GHz)

Residual Errors	SOLT (Fixed) (dB)	SOLT (Sliding) (dB)	SOLT (Offset) (dB)	TRL (dB)
Directivity	−40	−52	−60	−60
Match	−35	−41	−42	−60
Tracking	±0.1	±0.047	±0.035	±0

the most convenient, but least accurate, calibration procedures. The use of an offset load offers some improvement in accuracy over the sliding load. The TRL calibration process has the advantage of increased accuracy without added complexity. One disadvantage of the technique is that the calibration is relatively narrow band.

Another calibration technique, called line-reflect-match (LRM), can also be implemented for noncoaxial transmission media. LRM gives results similar to those of the TRL method. Studies indicate that for on-wafer tests the LRM calibration is more accurate than both the SOLT and TRL methods [5]. For a two-port measurement, all methods determine a twelve-term error model that quantifies errors due to impedance mismatches and signal leakage. The LRM technique is relatively broadband but requires a nearly ideal matched load to obtain accuracy.

A microwave probe station is typically calibrated at the probe tips using impedance standards amenable to probing. Therefore, precise error-corrected S-parameter measurements are easily made using a microwave probe station in an automated test configuration. More accurate device models can be attained by using probing techniques because effects due to packaging do not degrade the measurements. At chip or package level, fixturing is required to interface to the device. This requires the removal of the fixture effects on the measurements by means of a sophisticated de-embedding technique [6] or by calibrating with standards in the device test fixture. Both solutions are possible, but they are prone to error if care is not taken during either the calibration or de-embedding processes. The TRL and LRM calibration offer the advantageous ability to manufacture easily standards in transmission line media such as a microstrip.

3.2.2 Pulsed Measurements

In addition to S-parameter measurements, pulsed current-voltage measurements can be used in conjunction with a large-signal model parameter extraction scheme to determine model parameter values. For power applications, pulsed drain-source measurements can be used to determine the avalanche characteristics of the gate-drain diode without damaging the device. Similar measurements of the forward conduction phenomena can also be made.

By pulsing the gate at discrete voltage levels and measuring the drain current, a family of pulsed *I-V* curves is established. The device is pulsed in the active region for an amount of time that is short compared to the frequency at which the output conductance dispersion is occurring (typically below 1 MHz). A time on the order of 0.5 to 1.0 μs is acceptable for most devices.

Pulsed current-voltage measurements are considerably more difficult to make than dc measurements, but they do produce better predictions of harmonic distortion than models based on dc *I-V* curves [7]. The pulsed data produce predicted output resistance and transconductance values for a device that are comparable to the values determined using RF measurements. For purposes of obtaining harmonic distortion predictions, these values are much better than the values obtained from dc measurements. Figure 3.10 is a block diagram of a measurement system capable of obtaining pulsed gate measurements. The pulse generator is used as the gate-source bias. A pulse with a period of 1 μs is used to bias the device into the active region. After a short delay, the sample-and-hold (S/H) amplifier is triggered to sample the drain-source current via the current probe. The task of providing the proper timing for this type of measurement is nontrivial. The critical part of the timing sequence is triggering the S/H amplifier after the device has reached its "on"

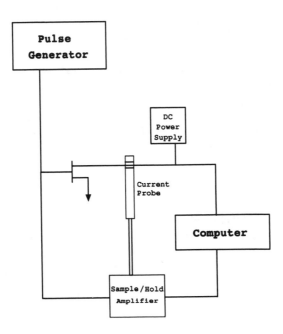

Figure 3.10 Measurement equipment configuration used to make pulsed gate current-voltage measurements of a MESFET or HEMT.

state and allowing the current probe enough response time. The device is then biased into cutoff to allow for the drain-source voltage to be set to the next value. This process is repeated for various pulse amplitudes until a set of *I-V* curves is produced.

Although this method has been used for parameter extraction purposes, it is relatively difficult to perform accurately. The measurement also provides information about the device under operating conditions that do not approximate actual CW operating conditions. An alternative, easier to apply, more accurate method to determine most model parameter values, which is valid at RF frequencies, involves the use of *S*-parameter data at multiple bias settings. The measurements required for this technique were described in the previous paragraphs. The parameter extraction algorithm for such a technique is described in detail in Chapter 4.

One area in which small-signal characterization techniques are not easily applied to the extraction of model parameters is for those parameters used to describe avalanche breakdown in the gate-drain region. Because high frequency avalanche breakdown characteristics are dramatically different from dc breakdown characteristics [8], a device biased such that dc avalanche is taking place is not approximating the behavior of the device during normal operating conditions. In normal operation, the device is dc biased well below the voltage levels needed to cause avalanche breakdown. The gate-drain breakdown then occurs only as large RF signal swings drive the voltage levels into avalanche conditions for short periods of the waveform cycle. The dc avalanche current are generally smaller than pulsed values and are not relevant to the RF performance of a device.

A pulsed measurement system can be used to measure these avalanche characteristics of the gate-drain barrier [9]. In this system, the drain-source voltage is pulsed while dc is applied to the gate-source terminal. Similar measurements can be made on-wafer using a microwave probe station. To obtain repeatable results from such measurements, as many as 500 measurements need to be averaged for every measurement point [10].

3.2.3 Low Frequency Dispersion Measurements

The measurement of the shift in output resistance can be made using the test configuration of Figure 3.11 [11]. Coaxial transmission lines are utilized for bias and signal paths to avoid unwanted device oscillations. The circuit is monitored for oscillations using a spectrum analyzer. A low frequency sine wave generator is used with signal voltage amplitudes set to minimal value (50 to 250 mV typically) to ensure the device is not overdriven. Measured small-signal output resistance as a function of frequency obtained from the measurement system of Figure 3.11 is presented in Section 1.4 (Figures 1.33 through 1.35). Similar equipment configurations can be utilized to measure low frequency dispersion in MESFET and HEMT transconductance characteristics.

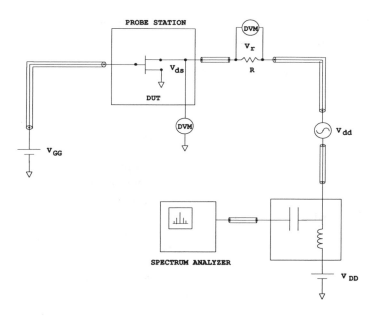

Figure 3.11 Measurement configuration used to characterize low frequency dispersion of output resistance in MESFETs or HEMTs.

3.3 LARGE-SIGNAL MEASUREMENTS

3.3.1 Load-Pull Measurements

One direct large-signal measurement technique designed to characterize device properties is the load-pull measurement. Load-pull characterization involves embedding the device to be tested into measurement circuitry that can be impedance tuned. The measurement system simultaneously monitors the tuned impedance of the characterization circuitry and the performance of the device. Device response is then recorded under the variable load conditions. The resulting loci of impedances required to obtain a constant performance parameter (i.e., output power, power added efficiency, *et cetera*) are typically displayed on a Smith chart in the form of closed contours. The load-pull contours are determined for one frequency at a time. For wideband characterization, load-pull measurements can be taken at several discrete frequencies within the band of interest. One difficulty with this type of characterization is that the impedances presented to the device at all frequencies (not just the fundamental) affect the performance. The load-pull contours also apply to only the incident power level utilized in the measurement.

Although load-pull measurements have been exploited for several years [12–21], the measurement equipment is still custom designed and expensive. Each system is unique, with advantages and shortcomings over other reported systems. Although many variations of the load-pull measurement apparatus exist, the equipment configurations can be classified into two general categories. *Traditional* load-pull systems utilize mechanical adjustments to achieve the required circuit impedance tuning. Slide-screw tuners [15], dielectric slug tuned air lines [12] and microstrip tuning [13] are among the techniques that can be used for this purpose. An alternative to the traditional load-pull technique is the *active-load* method. This method uses two phase- and amplitude-controlled RF signals applied to the device. No physical impedance tuner is required for this approach.

Figure 3.12 illustrates a typical equipment configuration that can be used to realize a traditional load-pull system. A large amplitude signal is supplied to the DUT from the traveling wave tube (TWT) amplifier. The input tuner is used to optimize the input match and to assure maximum power transfer to the device. The output tuner is tuned to realize some specific performance criteria. Device performance and the impedance seen by the device are determined by the network analyzer. Bias to the device under test is supplied either via the measurement fixture or through optional bias tees placed in the system. Automation of such systems usually involves control of servomotors that mechanically tune the input and output tuners. Modifications to this kind of measurement system can be implemented to allow measurement and independent tuning of the harmonic content of the signal as well as the output power in the fundamental frequency [14].

The traditional load-pull method has some limitations related to the output tuner. Accurate measurements depend on the precision and calibration of an

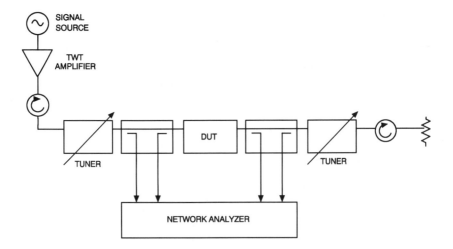

Figure 3.12 Typical test equipment configuration used to make traditional large-signal load-pull measurements.

impedance tuner. Such tuners are difficult to obtain at high frequencies (especially above approximately 12 to 18 GHz). In addition to precision repeatability and accurate calibration, the tuner also needs to be capable of presenting a large range of reflection coefficients to the device. Ideally, the tuner can present a reflection coefficient with magnitude varying from 0 to 1 and at any desirable angle. Loss in the tuner, however, limits the realizable reflection coefficient of the measurement system. As an example, a loss as little as 1 dB in the tuner reduces the realizable reflection coefficient magnitude to about 0.8. Additional losses in isolators, bias tees, cables, *et cetera,* further reduce this figure. Such problems can become extremely limiting at high frequencies.

Active-load measurement systems overcome many of the problems associated with traditional load-pull systems. Figure 3.13 presents a diagram of an active-load measurement configuration. The major advantage of this type of system is that the

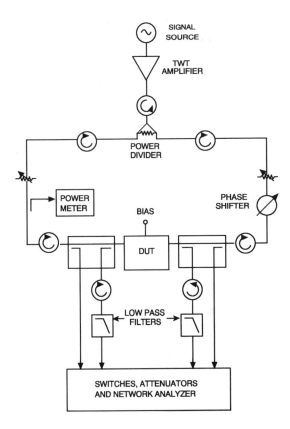

Figure 3.13 Typical test equipment configuration used to make active-load load-pull measurements.

load reflection coefficient obtained is not limited in magnitude. By independently adjusting the input attenuators in the system, virtually any reflection coefficient can be synthesized. The system can be automated to obtain output power as well as third-order intercept point contours [17]. Other systems utilizing a similar equipment configuration have also been employed successfully [18–21]. The measurement configuration of Figure 3.13 is completely symmetric and offers an advantage over similar systems in that the device under test sees a 50-Ω matched system at all harmonics of the applied signal.

In the test system, a large-amplitude signal is provided by the TWT and is divided into two signals that drive each port of the DUT. The power level and relative phase of the two waves incident on opposite ports of the device are controlled with two variable attenuators and a phase shifter. Important features of the measurement system are the two bidirectional couplers, switches, and filters. These components serve to precondition the high level signals before they reach the network analyzer. The low-pass filters serve to present the characteristic impedance to the harmonic components of the reflected waves. This feature is not common to most load-pull systems. Isolators are used throughout the system to protect sensitive components from the high power levels that are present.

Calibration of active-load systems is often accomplished using conventional network analyzer standards at reduced power levels [17, 19, 21]. Short and through standards are useful in determining offset electrical lengths, while further calibration is obtained using open and resistive terminations.

Large-signal measurements are performed after calibration. Loads are adjusted using tuners for traditional systems or using attenuators and phase shifters for active-load systems. Systems can be designed to search for the tuning conditions that produce a desired performance [12] or that record performance at a number of prescribed impedances [15]. When the latter approach is used, interpolation schemes are required to produce performance contours. The accuracy of these systems is limited primarily by the accuracy with which the terminations and RF power levels can be determined. Unknown component losses, limited coupler directivity, and connector mismatch are all sources of error.

Figure 3.14 presents load-pull data for a microwave MESFET [21]. A signal frequency of 26 GHz was used for this measurement. The point at the center of the contours represents the impedance, which when presented to the output of the device results in maximum power delivered to the load for this input power level. Each contour line represents the loci of reflection coefficients which result in a constant output power level. The contours of the figure are spaced 1 dB apart. Similar contours can be generated with a load-pull system to specify power-added efficiency, power gain, or third-order intercept point performance. Data taken at another frequency or input power level will result in contours different from those illustrated.

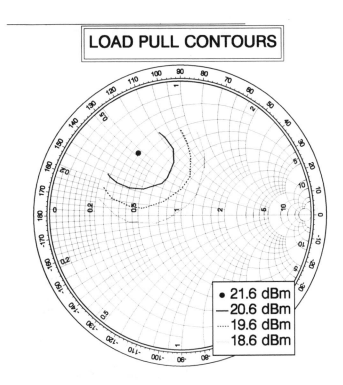

Figure 3.14 Measured results from a load-pull measurement of a microwave MESFET at 26 GHz [21]. Reprinted by permission (©1987 IEEE).

The load-pull measurement results can be used directly in the design of power amplifiers. The data can also be used to advantage when choosing the optimal devices for specific power applications or to verify large-signal models. Load-pull measurements, however, do not provide an easy means for performing parameter extraction of large-signal model parameters. Not only do the measurements require a significant amount of expensive microwave equipment and considerable characterization expertise, but simulating the results by using CAD tools to perform parameter extraction is also computationally intensive.

Another technique that has been used for large-signal device characterization is measurement of large-signal *S*-parameters. The characterization procedure is similar to conventional *S*-parameter measurements, but is implemented under conditions of high RF signal levels. As in the case of load-pull measurements, the data obtained in this manner are applicable only at the power levels used to perform the experiment. The resulting large-signal *S*-parameter data are closely related to load-pull data [16]. The limitations and capabilities of the two techniques are, therefore, very similar.

3.3.2 Two-Tone Harmonic Content Measurements

An alternative direct measurement of nonlinear device behavior is a two-tone harmonic content measurement. This measurement can be simpler to make and involve less expensive configurations of test equipment than load-pull measurements. To make the measurement, the output spectral content of a device is recorded while two input signals, closely spaced in frequency and of equal magnitude, are applied to the input of the device. The measurement is made for increasing input power levels. Although tuning can be used to optimize performance, no tuning or variable phase-shift elements are required when the measurement is being used to validate models and parameter extraction techniques. The measurement is typically made either on-wafer using coplanar wafer probes or in a standard 50-Ω test fixture. Results are typically displayed as x-y plots of output power for each frequency component of the output signal (harmonic) as a function of input signal power. As in the case of load-pull measurements, the data apply only to the frequency used for the measurement.

Figure 3.15 presents a diagram of an on-wafer two-tone harmonic measurement configuration. The two input signals, F_1 and F_2, are closely spaced in frequency (typically on the order of 2 to 20 MHz apart). Isolators are used with the signal sources to ensure a pure spectral response for the input waveform. Without good isolation, ensuring that the sources do not affect each other is difficult. The signals are power combined and applied to the device. A spectrum analyzer is used to measure the magnitude of the spectral components in the output waveform. Attenuators are useful to reduce the effect of mismatches in the equipment config-

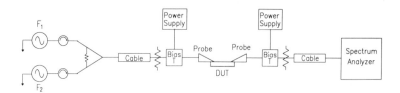

Third–Order Intercept Point Measurement Configuration

Figure 3.15 Measurement equipment configuration used to make on-wafer two-tone harmonic content measurements of microwave devices.

uration and bias tees may be used to apply bias to the device under test. For some high power devices, additional amplifiers may be required to boost the level of the input signals.

The only calibration required for the system is the determination of losses in the signal path. This is done for both the input and output portions of the measurement system. A short through line is useful for determining actual probe losses.

To make the measurement, signal sources F_1 and F_2 are set to a specific input power level. The signals should be of equal magnitude. Magnitudes for fundamental second- and third-order mixing products of the two input signals are determined from the spectrum analyzer. Once the input and output power levels have been recorded, input signal levels are increased and the measurement repeated. Initially, power levels are typically set very low and then increased to power levels slightly beyond a 1-dB compression of the device gain.

Figure 3.16 presents results of a two-tone harmonic content measurement for a microwave MESFET. The equal power signals were at frequencies of 2.10 and 2.12 GHz for this measurement. The input power level specified on the x-axis is the power of only one input signal. The total power is twice this, or 3 dB higher. Figures 3.17 and 3.18 plot the results of the same measurement in terms of power

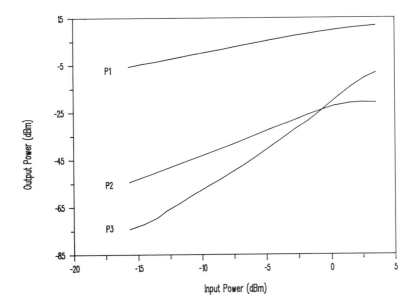

Figure 3.16 Measured results from two-tone harmonic content measurements using the test system illustrated in Figure 3.15. The plot displays measured output power in the fundamental (P1), second (P2), and third (P3) harmonic as a function of input power for a microwave MESFET; signal sources were operated at 2.10 and 2.12 GHz.

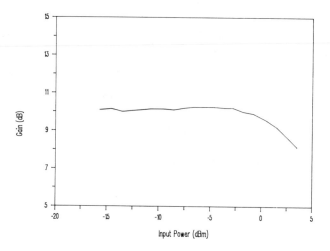

Figure 3.17 Measured results from two-tone harmonic content measurements using the test system illustrated in Figure 3.15. The plot displays measured power gain as a function of input power for a microwave MESFET.

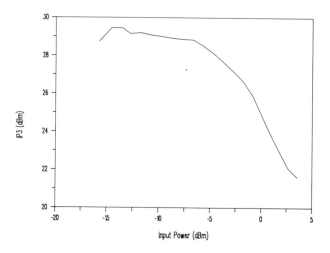

Figure 3.18 Measured results from two-tone harmonic content measurements using the test system illustrated in Figure 3.15. The plot displays the measured third-order intercept point as a function of input power for a microwave MESFET.

gain and third-order intercept point, respectively. The third-order intercept point is a figure of merit that is indicative of the dynamic range capability of a device or circuit [22]. For this measurement, input power levels increased to approximately 2 dB of gain compression were observed.

The information obtained from the two-tone measurement is not directly applicable in circuit design applications. This information can be useful, however, for verification of device model accuracy or of the validity of a parameter extraction technique. The amount of computer simulation required to predict the two-tone response of a device in a 50-Ω system is far less than that required to predict load-pull contours. Thus, the two-tone measurement made in a 50-Ω system represents a fast and efficient method of model verification and parameter extraction technique validation.

3.4 NOISE CHARACTERIZATION

To utilize empirical noise models for a MESFET or HEMT, certain electrical measurements are required that provide noise information about the device. For the empirical noise model proposed by Gupta (described in Section 2.2.1), the noise current spectral density is needed. This current is determined indirectly by first measuring the noise power spectral density at the drain-source terminals with the gate-source terminals short circuited. Then, given the output resistance r_{ds} of the device and the impedance of the measurement equipment R, the current is computed from equation (2.54). Because this model requires only a power measurement, the test can be made with relatively inexpensive equipment and very little test time. This is especially convenient for making measurements at multiple device bias conditions or frequencies and for making on-wafer measurements. The empirical noise model proposed by Fukui in Section 2.2.1 requires a measurement of the device noise parameters Γ_{opt}, R_n, and F_{min}. The direct measurement of these parameters requires much more sophisticated equipment. Also, the measurement is time consuming, requires calibrations to characterize the equipment, and thus requires automation for most applications.

For physically based models, measurements of device dimensions and material properties are required. Difficulties with direct measurement of many of these properties are discussed in Section 4.3.5. Alternatives to such measurements are also discussed.

3.4.1 Spectral Density Measurements

Equipment required to measure noise power spectral density (P_{out}) at the drain-source terminal of a MESFET or HEMT is shown in Figure 3.19. The measurement apparatus consists of a bandpass filter, an amplifier, and a power meter or

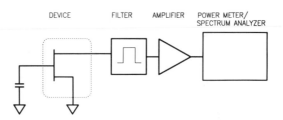

Figure 3.19 Block diagram of a measurement system to measure the noise power at the device drain-source terminals. A high gain, low noise amplifier is often needed to raise the noise to a level where it can be measured easily.

spectrum analyzer to measure noise power. Connection to the drain-source terminals is assumed to be via an RF probe. In some cases, the spectrum analyzer is capable of defining an effective noise bandwidth and thus a filter is not required.

Because the noise power density at the device drain-source terminals is very small (typically 2×10^{-20} W/Hz), a high gain amplifier is often required to increase the power to a level that is easily measured by the power meter or spectrum analyzer. Also, the amplifier generates internal noise. Consequently, a portion of power measured by the analyzer is due to the amplifier and must be analytically removed to determine P_{out}. For these reasons, and also to improve measurement accuracy, the amplifier should exhibit a low noise figure. This ensures that the measured noise power is due primarily to the device. The amplifier requirements are illustrated later in this section.

One of the key factors influencing this measurement is the effective noise bandwidth, B_N. Because the spectral content of the noise is white, the power measured is proportional to the bandwidth. This bandwidth must be precisely defined, either by the spectrum analyzer or by a bandpass filter. Even when the analyzer is used as a bandpass element, a correction factor is sometimes required to convert the analyzer bandwidth to the effective noise bandwidth.

When a bandpass filter is utilized, the effective noise bandwidth is defined as that which is equivalent to an ideal filter of infinite order which would pass the same amount of white noise [23]. Any realizable filter has finite order and cannot exhibit a rectangular passband response. However, as the filter order increases, the bandwidth approaches the effective noise bandwidth. For example, a fifth-order Butterworth filter has an effective noise bandwidth that is about 2% greater than the 3-dB bandwidth. Considering only filter order, the ratio of effective noise bandwidth to 3-dB bandwidth for a Butterworth filter varies from 1.57 (first order) to close to 1 at higher order. For filters with different responses, such as Chebychev, the noise bandwidth may be greater or less (depending on filter order and passband ripple) than the half-power bandwidth. A tabulation of effective noise bandwidths for several filter types and orders is presented in Reference [23].

The noise power measured by the spectrum analyzer or power meter (Figure 3.19) includes noise generated by the device and amplifier. Also, the noise power is affected by losses in the filter, RF probe, interconnecting cables, and impedance mismatches. The adverse effects due to any impedance mismatch are significantly reduced by adding an isolator in close proximity to the device drain-source terminals. The noise generated by these external components, however, must be characterized to allow removal of their contribution from the measurement. When the external components between the device-drain source terminals and the meter are considered as a cascaded network, a separate measurement that determines the gain and noise figure of this network may be a sufficient calibration. Alternatively, the gain and noise figure of each component can be measured and the cascaded gain and noise figure computed from

$$F_{cas} = F_{probe} + (F_{amp} - 1)/G_{probe} + (F_{filter} - 1)/(G_{probe}G_{amp}) + \cdots \qquad (3.5)$$

$$G_{cas}\,(dB) = G_{probe} + G_{amp}\,(dB) + G_{filter}\,(dB) + \cdots \qquad (3.6)$$

where F_{cas} and G_{cas} are the cascaded noise figure and gain, respectively, of the external circuitry (RF probe, amplifier, interconnecting cable, *et cetera*).

Removal of the noise contribution due to the external components from power measured at the analyzer is accomplished using

$$P_{out} = P_{meas}/(F_{cas}G_{cas}) \qquad (3.7)$$

where P_{out} is the noise power at the device drain-source terminals and P_{meas} is noise power measured at the power meter.

Equations (3.5) through (3.7) allow computation of the power spectral density at the device drain-source terminals to be accomplished. However, the power spectral density P_{out} must be normalized to a 1-Hz bandwidth given the effective noise bandwidth B_N of the measurement.

Equation (3.5) indicates the amount of noise power due to the device, amplifier, and filter that is measured at the analyzer. For an accurate measurement, the measured noise power (P_{meas}) should be primarily due to the device (P_{out}), which is amplified by the external components ($P_{out}G_{cas}$). This requires a high gain amplifier with a low noise figure. The following example illustrates the noise requirements of the external circuitry.

Table 3.2 presents noise power spectral measurements of a 0.5×300-μm MESFET using the measurement technique shown in Figure 3.19. For this measurement, an isolator is connected near the device drain-source terminal to reduce impedance mismatch effects. The external components are characterized with a resulting gain and noise figure of 36.1 and 3.2 dB, respectively. From equation (3.7), the noise produced by the device is computed and shown in Table 3.2. All measurements are normalized to a 1-Hz bandwidth.

Table 3.2
Noise Power Spectral Density Measurements of a $0.5 \times 300\text{-}\mu m$ MESFET

I_{ds}/I_{dss}	P_{meas} (dBm/Hz)	P_{out} (W/Hz)	$P_{out}G_{cas}$ (dBm/Hz)
0.10	-128.0	1.86×10^{-20}	-131.2
0.20	-128.0	1.86×10^{-20}	-131.2
0.30	-126.0	2.95×10^{-20}	-129.2
0.40	-125.0	3.72×10^{-20}	-128.2
0.50	-124.0	4.68×10^{-20}	-127.2
0.60	-123.0	5.89×10^{-20}	-126.2
0.70	-121.6	8.13×10^{-20}	-124.8

A comparison of the noise power due to the device ($P_{out}G_{cas}$) indicates that it generates approximately one-half of the measured noise. The remaining noise power is due to the amplifier, filter, probe, *et cetera*. For the highest accuracy, the noise generated by the external components should be limited to a small fraction of the measured power to ensure a reliable computation of the device noise power spectral density. This is accomplished when the cascaded noise figure F_{cas} is low— typically below 3 dB.

3.4.2 Fully Automated Minimum Noise Figure Measurements

The Fukui model described in Section 2.2.1 requires knowledge of noise parameters Γ_{opt}, F_{min}, and R_n. These parameters must be known at a minimum of one frequency and at the desired device bias. From these data, the model predicts noise parameters as a function of frequency for the specified device bias. To utilize the model, therefore, a measurement of device noise parameters is required.

As previously discussed in Chapter 2 the device noise figure varies depending on the admittance connected at the device input according to

$$F(Y_g) = F_{min} + \frac{R_n}{G_g}[(G_g - G_{opt})^2 + (B_g - B_{opt})^2] \tag{3.8}$$

where

F_{min} = minimum value of F with respect to Y_g,
R_n = equivalent noise resistance, and
Y_g = admittance connected at the device input.

Note that $Y_{opt} = G_{opt} + jB_{opt}$ is the admittance value at which $F = F_{min}$.

One method of directly measuring Y_{opt}, F_{min}, and R_n, is to monitor the device noise figure while adjusting the source admittance for minimum noise (Figure 3.20). At this point, the measured noise figure is the minimum device noise figure (F_{min}) and the source admittance is Y_{opt}. Also, R_n can be determined by solving equation (3.8). Hence, Y_{opt}, F_{min}, and R_n are determined by direct measurement (assuming the measurement equipment, cabling, and fixturing are properly characterized and their effects removed from the measurement).

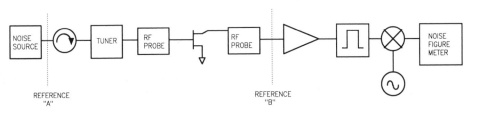

Figure 3.20 Block diagram of a system to directly or indirectly measure device noise parameters F_{min}, R_n, and Y_{opt}.

While this direct method is at least, in principle, mathematically correct and even practical, it suffers limitations, particularly in accuracy. Searching for a minimum noise figure can be inaccurate because the minimum may be shallow and thus difficult to precisely detect. Also, this method places demanding requirements on the tuner, which must exhibit a wide range of admittance values. Further, in automated systems, the tuner is often digitally controlled and exhibits only discrete states—none of which may necessarily correspond to the one required. Thus, the accuracy of this method is limited by equipment limitations and the requirement to detect precisely a minimum device noise figure.

Another method of measuring these noise parameters that overcomes these limitations and significantly improves accuracy utilizes an indirect approach [24–27]. The device noise figure is measured at four or more unique source impedances, none of which needs to correspond to Y_{opt}. These measured noise figures (F_1, F_2, F_3, F_4, et cetera) at source admittances (Y_1, Y_2, Y_3, Y_4, et cetera) along with equation (3.8) allow an algebraic solution for Y_{opt}, F_{min}, and R_n to be obtained. When more than four measurements are utilized, error correction methods can be included to minimize systematic measurement error and further improve accuracy [26, 27].

As mentioned, numerous sources of error contribute uncertainty in noise measurements, particularly in measuring the minimum noise figure F_{min}. This is often overlooked, especially when examining published MESFET or HEMT noise parameters that are shown as well behaved smooth monotonic functions of frequency. These smoothed data are the result of functionally fitting the measured data to some criteria—such as a least-squares error. Because noise measurements utilize sensitive equipment to measure power, measurement repeatability and operator variability also affect the results. Although this limitation is equipment, device, and level dependent, some studies have suggested a ± 0.2-dB uncertainty in measuring the minimum noise figure of low noise MESFETs biased at low current levels [28].

In addition to these effects, noise generated by other sources external to the equipment, such as other test equipment in close proximity, may further affect noise measurements. These effects are additive and result in considerable scatter in the measured results. As an illustration, noise parameters for a low noise device were measured using the indirect method based on nine unique source admittance settings. A minimum noise figure, R_{opt}, X_{opt}, and R_n were measured from 2 to 18 GHz. A plot of the measured results and an analytical functional fit to the data are shown in Figure 3.21. As illustrated, measurement uncertainty limits accuracy, particularly for low noise devices. This implies considerable uncertainty in measuring noise figures below perhaps 0.3 dB.

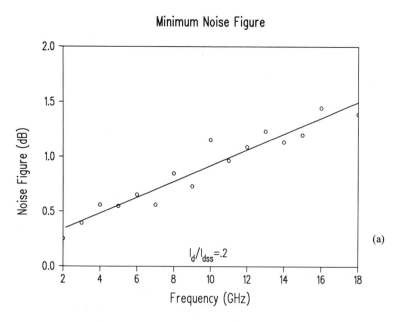

Figure 3.21 Measured noise parameters using the indirect method at nine unique source admittances exhibit considerable scatter. Both raw data and a functional fit to the data are shown: (a) minimum noise figure; (b) R_n; (c) R_{opt}; (d) X_{opt}.

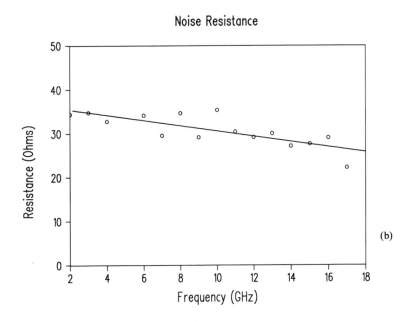

(b)

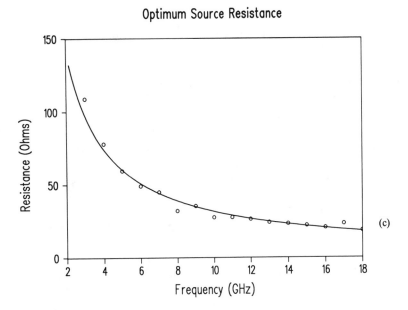

(c)

Figure 3.21 continued

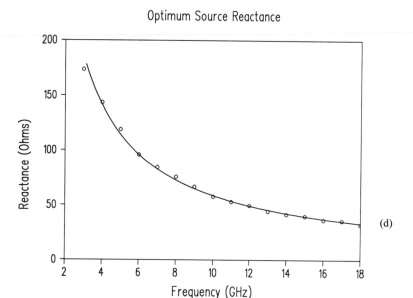

Figure 3.21 continued

At this point, some remarks regarding the measurement equipment, characterization, and calibration are in order. The basic equipment for either indirect or direct noise measurement is shown in Figure 3.20. The noise source generates noise powers at two levels corresponding to noise temperatures (T_1 and T_2). Based on these known levels, a noise meter detects noise power and determines the noise figure for the circuitry between the source and meter. Depending on frequency, application, and equipment limitations, an amplifier, bandpass filter, and mixer may be required. In some situations, these components are needed to down convert the noise to a frequency range suitable to the meter. The isolator preceding the amplifier maintains a constant source impedance at the input of the LNA despite changes in device characteristics. This is necessary because changes in the tuner impedance will otherwise affect the input impedance to the LNA and thus change its noise figure. In addition to these components, the tuner provides several source impedance states and the RF probes allow contact with the FET or HEMT.

For calibration purposes, two reference planes "A" and "B" (Figure 3.20) can be defined. All components to the right of plane B are considered to be part of the meter and are calibrated as part of the meter. With this definition, the equipment measures the cascaded noise figure (F_{cas}) of the components between planes A and B. Noise contributions due to the external equipment must be characterized and removed to yield the device noise figure.

The circuitry to the left and right of the device can be independently characterized in terms of gain and noise figure (G_1, F_1) and (G_2, F_2), respectively. Thus, the cascaded noise figure (at planes A and B) is given by

$$F_{cas} = F_1 + \frac{F_{device} - 1}{G_1} + \frac{F_2 - 1}{G_1 G_{device}} \tag{3.9}$$

Thus, the FET noise figure (F_{device}) is determined from equation (3.9).

In addition to this equipment calibration, effects due to impedance mismatches can also be corrected based on measured S-parameters [25, 29].

The power of this technique is that fully automated measurements of device noise parameters can be made accurately. The test equipment is calibrated and systematic error removed from the results.

REFERENCES

[1] H. Fukui, "Determination of the Basic Device Parameters of a GaAs MESFET," *Bell Syst. Tech. J.,* Vol. 58, March 1979, pp. 771–797.

[2] H. Willing, C. Rauscher, and P. de Santis, "A Technique for Predicting Large-Signal Performance of a GaAs MESFET," *IEEE Trans. Microwave Theory Tech.,* Vol. MTT-26, December 1978, pp. 1017–1023.

[3] "Network Analysis—Applying the HP 8510B TRL Calibration for Non-Coaxial Measurements," Hewlett-Packard Product Note 8510-8, October 1987.

[4] "Vector Measurements of High Frequency Networks," Hewlett-Packard Seminar Notes, April 1989.

[5] S. Lautzenhiser, A. Davidson, and K. Jones, "Improve Accuracy of On-Wafer Tests Via LRM Calibration," *Microwaves & RF J.,* January 1990, pp. 105–109.

[6] J. Staudinger, "A Two-Tier Method of De-Embedding Device Scattering Parameters Using Novel Techniques," MS Thesis, Arizona State University, May 1987.

[7] M. Paagi, P.H. Williams, and J.M. Borrego, "Nonlinear GaAs MESFET Modeling Using Pulsed Gate Measurements," *IEEE Microwave Theory Tech. Symp. Digest,* 1988, pp. 229–231.

[8] S.H. Wemple, W.C. Niehaus, H.M. Cox, J.V. DiLorenzo, and W.O. Schlosser, "Control of Gate-Drain Avalanche in GaAs MESFETS," *IEEE Trans. Electron Devices,* Vol. ED-278, June 1980, pp. 1013*ff.*

[9] S.H. Wemple, M.L. Steinberger, and W.O. Schlosser, "Relationship Between Power Added Efficiency and Gate-Drain Avalanche in GaAs MESFETS," *Electron. Lett.,* Vol. 16, June 5, 1980, pp. 459–460.

[10] J.F. Vidalou, F. Grossier, M. Camiade, and J. Obregon, "On-Wafer Large Signal Pulsed Measurements," *IEEE Microwave Theory Tech. Symp. Digest,* 1989, pp. 831–834.

[11] J.M. Golio, M.G. Miller, G.N. Maracas, and D.A. Johnson, "Frequency Dependent Electrical Characteristics of GaAs MESFETS," *IEEE Trans. Electron Devices,* Vol. ED-37, May 1990, pp. 1217–1227.

[12] J.M. Cusack, S.M. Perlow, and B.S. Perlman, "Automatic Load Contour Mapping for Microwave Power Transistors," *IEEE Trans. Microwave Theory Tech.,* Vol. MTT-22, December 1974, pp. 1146–1152.

[13] H. Abe and Y. Aono, "11 GHz GaAs Power MESFET Load-Pull Measurements Utilizing a New Method of Determining Tuner Y-Parameters," *IEEE Trans. Microwave Theory Tech.,* Vol. MTT-27, May 1979, pp. 394–399.

[14] R.B. Stancliff and D. Poulin, "Harmonic Load-Pull," *IEEE Microwave Theory Tech. Symp. Digest,* 1979, pp. 185–187.

[15] M. Pieroint, R.D. Pollard, and J.R. Richardson, "An Automated Measurement Technique for Measuring Amplifier Load-Pull and Verifying Large-Signal Device Models," *IEEE Microwave Theory Tech. Symp. Digest,* 1986, pp. 625–628.

[16] R. Tucker, "RF Characterization of Microwave Power FETs," *IEEE Trans. Microwave Theory Tech.,* Vol. MTT-29, August 1981, pp. 776–781.

[17] J.M. Nebus, J.P. Villotte, J.F. Vidalou, L. Hagerman, H. Jallageas, and M.C. Albuquerqu "Optimized C.A.D. of Power Amplifiers, for Maximum Added Power or Minimum Third Order Intermodulation, Using an Optimization Software Coupled to a Single Tone Source and Load Pull Set-Up," *IEEE Microwave Theory Tech. Symp. Digest,* 1988, pp. 1049–1052.

[18] Y. Takayama, "A New Load-Pull Characterization Method for Microwave Power Transistors," *IEEE Microwave Theory Tech. Symp. Digest,* 1976, pp. 218–220.

[19] R.S. Tucker and P.D. Bradley, "Computer-Aided Error Correction of Large-Signal Load-Pull Measurements," *IEEE Trans. Microwave Theory Tech.,* Vol. MTT-32, March 1984, pp. 296–300.

[20] M. Lajugie, F. Grossier, A. Silbermann, and Y. Bender, "Full Characterization of GaAs Power MESFET and Accurate Load-Pull Contours Prediction," *IEEE Microwave Theory Tech. Symp. Digest,* 1986, pp. 339–342.

[21] K. Kotzebue, T.S. Tan, and D. McQuate, "An 18 to 26.5 GHz Waveguide Load-Pull System Using Active-Load Tuning," *IEEE Microwave Theory Tech. Symp. Digest,* 1987, pp. 453–456.

[22] T.S. Laverghetta, *Microwave Measurements and Testing,* Norwood, MA: Artech House, 1976.

[23] R.D. Shelton and A.F. Adkin, "Noise Bandwidth of Common Filters," *IEEE Trans. Communication Technol.,* December 1970, pp. 828–830.

[24] R. Froelich, "Measurement of GaAs FET Noise Parameters," Watkins-Johnson Company Technical Notes, Vol. 13, No. 6, November/December 1986.

[25] R. Froelich, "Automated Noise-Parameter Measurements Using a Microwave Probe," Watkins-Johnson Company Technical Notes, Vol. 16, No. 1, January/February 1989.

[26] R.Q. Lane "The Determination of Device Noise Parameters," *Proc. IEEE,* Vol. 57, August 1969, pp. 1461–1462.

[27] M. Mitama and H. Katoh, "An Improved Computational Method for Noise Parameter Measurement," *IEEE Trans. Microwave Theory Tech.,* Vol. MTT-27, June 1979, pp. 612–615.

[28] R. Lucero, C. Moyer, R. Viatkus, and M. Dydyk, "Noise Characterization Uncertainty of Microwave Devices Under Low Current Operation," *IEEE Microwave Theory Tech. Symp. Digest,* Vol. III, 1989, pp. 893–896.

[29] J.S. Gedymin, "Rid Noise from Tests of Unstable Transistors," *Microwaves & RF J.,* October 1986, pp. 113–116.

Chapter 4
PARAMETER EXTRACTION

Before any device model can be used, values for the parameters required of the model must be determined. The process of determining these parameters is called *parameter extraction.* As discussed in the introductory comments of Chapter 3, this process must be closely related to both the characterization data taken and the actual model used. Figure 4.1 illustrates the general relationship between device characterization, parameter extraction, device modeling, and circuit simulation for one approach to the microwave circuit design problem. Other approaches involving different characterization and parameter extraction schemes may also be used. The processes described by Figure 4.1 apply to small-signal, large-signal, and noise modeling applications (the small-signal equivalent circuit models are determined as an intermediate step toward large-signal model determination). Noise model parameters also require some small-signal modeling information. This connection between small- and large-signal parameter extraction is not common to all parameter extraction schemes.

Despite the fact that parameter extraction is as important to usable design procedures as modeling and characterization, this area of work has been largely neglected by the device modeling community. An effective engineering solution to the parameter extraction problem must be simple to implement, fast, and affordable.

This chapter presents algorithms that can be used with some of the characterization techniques presented in Chapter 3 and the models of Chapter 2 to extract parameters accurately and efficiently for small-signal, large-signal, and noise models. The algorithms presented apply to the technique illustrated in Figure 4.1, but can be generalized easily to apply to other characterization data or models not discussed in Chapter 2.

4.1 DIRECT CURRENT DATA EXTRACTION

Although dc measurements alone are not sufficient to describe the RF behavior of MESFETs and HEMTs, these measurements are easily performed and provide

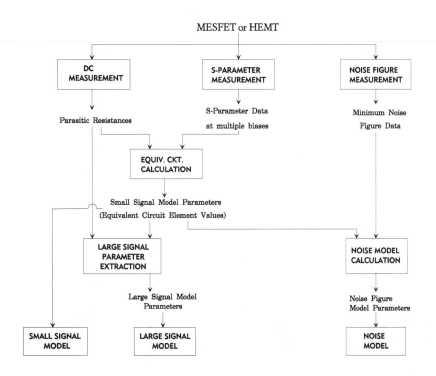

Figure 4.1 Illustration of the relationships between characterization, parameter extraction, and modeling of microwave MESFETs and HEMTs.

information that may be used to determine some of the important device parameters. Figure 4.2 illustrates the dc equivalent circuit for a MESFET or HEMT. The dc data may be used to extract the dc output conductance, dc transconductance, and the parasitic resistances. In addition, the forward-bias characteristics and dc breakdown properties of the gate-source and gate-drain diodes may be evaluated.

4.1.1 Current-Voltage Behavior

Several empirical and physically based models are discussed in Chapter 2. The large-signal models relate the drain-source current and device capacitances to the bias voltages. Although dc measured data are not easily applied to device capacitance relationships, they can be related directly to device current. One method of extracting the parameter values required of these models is to apply optimization of the modeled values of I_{ds} to the measured values at several bias points. The opti-

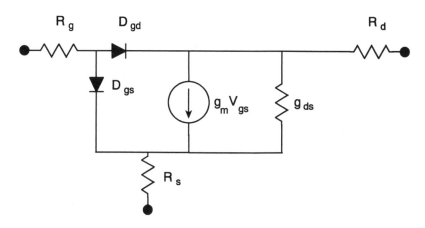

Figure 4.2 A dc model of a MESFET or HEMT.

mization process involves the minimization of a specified error function by adjustment of the model parameter values. One such error function that can be used in the optimization of the model to dc data is:

$$E_{Ids} = \sum_{i=1}^{N} \frac{(I_{modi} - I_{measi})^2}{I_{measi}^2} \qquad (4.1)$$

where I_{modi} is the modeled drain-source current at bias point i, I_{measi} is the measured current at bias point i, and N is the number of bias points considered.

Figures 4.3(a) and 4.3(b) illustrate the capability to match the advanced Curtice model of Chapter 2 to the measured dc behavior of a 0.5×300-μm MESFET. The optimization process used to produce these data is described in Section 4.3. The resulting accuracy is ultimately limited by the ability of the particular model employed to describe the device characteristics. The relative accuracy of several large-signal models is presented in Chapter 2.

Two other parameters that may be of interest are the output conductance and the transconductance of the device. The output conductance of the device is given by the derivative of I_{ds} with respect to V_{ds}, as expressed in equation (1.20). Similarly, the transconductance may be found by differentiating I_{ds} with respect to V_{gs}, as expressed in equation (1.21).

Optimization of the model to the measured drain-source current indirectly affects G_{ds} and G_m values. For analog applications, however, these parameters are of particular interest. Separate error terms similar to equation (4.1) can be formulated, with the total error term given by:

$$E_{tot}^2 = E_{Ids}^2 + E_{Gds}^2 + E_{Gm}^2 \qquad (4.2)$$

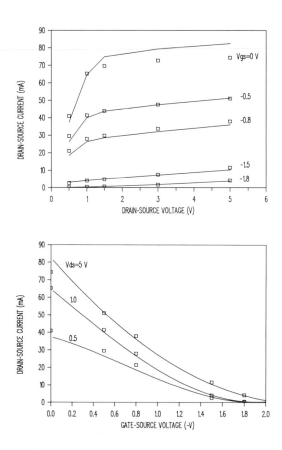

Figure 4.3 Comparison of predicted (solid line) and measured (squares) dc performance of a 0.5×300-
μm MESFET. The model used is the advanced Curtice model: (a) drain-source current as a
function of drain-source bias; (b) drain-source current as a function of gate-source bias.

The obvious advantage of using the dc data is the ease with which the measurements may be made. However, investigations have shown that dc characterization
in itself does not provide sufficient information to describe accurately the RF
behavior of the device [1, 2]. As described in Chapter 1, the output conductance
and transconductance of these devices are dispersive, thus models optimized to
only dc data do not accurately describe the RF performance. Additionally, because
the output conductance and the transconductance have significant impact on the
performance of most microwave circuits, these dc extracted models are not acceptable for use in the design of microwave circuits.

A solution to this problem will be described in Section 4.3. By using a combination of dc and RF measurements, accurate models can be developed.

4.1.2 Parasitic Resistances

Determination of the parasitic resistances of the MESFET and HEMT is an important step in the parameter extraction process for small-signal, large-signal, and noise models. Independent determination of these resistance values reduces the number of element values requiring optimization. This reduction makes the small-signal optimization process faster and more reliable. In addition, by improving the accuracy of the small-signal models, the ability to extract large-signal models accurately is improved. The parasitic resistance values also contribute significantly to the device's noise figure. Thus, accurate determination of these resistances is crucial to the success of all noise models.

Using the technique described in Section 3.1, the parasitic resistances of the gate, drain, and source of a MESFET may be determined by measuring the gate current as a function of gate voltage for the configuration shown in Figure 4.4. As described in Chapter 3, the measurements are made under three conditions: (a) source grounded, (b) drain grounded, and (c) both source and drain grounded. The series resistance values obtained for these three configurations may be related to the parasitic resistances as follows:

$$R_{ser} = R_a = R_g + R_s \tag{4.3a}$$

$$R_{ser} = R_b = R_g + R_d \tag{4.3b}$$

$$R_{ser} = R_c = R_g + R_d R_s / (R_d + R_s) \tag{4.3c}$$

These equations may be solved for R_g, R_d, and R_s in terms of R_a, R_b, and R_c as follows:

$$R_g = R_c - [R_c^2 - R_c(R_a + R_b) + R_a R_b]^{1/2} \tag{4.4a}$$

$$R_d = R_b - R_g \tag{4.4b}$$

$$R_s = R_a - R_g \tag{4.4c}$$

The values of the parasitic resistances determined in this way can be slightly sensitive to the bias conditions, as demonstrated in Figure 4.5. These results show that the source and drain resistance values for the 0.5×300-μm MESFET are fairly constant, except for a decrease at lower gate voltages. At these low forward gate bias levels, the measurement accuracy is reduced because the differences between measured and ideal current are small. The observed drop in resistance is attributable to this decrease in experimental accuracy. The gate resistance shows much greater dependence on V_g. This bias dependence is not typically included in microwave device models, making it difficult to specify a unique value of resistance from the dc measurement. In some circumstances, the value of gate resistance obtained from

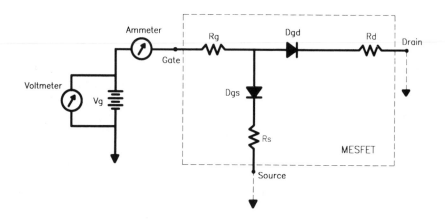

Figure 4.4 Measurement setup for determining parasitic resistances.

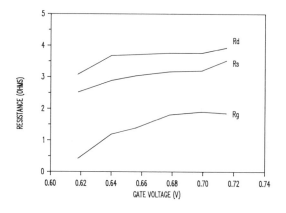

Figure 4.5 Drain, source, and gate resistance of a 0.5×300-μm MESFET as a function of gate voltage.

these measurements may not be acceptable for RF device modeling. Experience indicates that the best choice for a value of modeled gate resistance is often 10 to 50% higher than a value determined using this measurement method.

One criticism of this method is that the resistance values are obtained under forward-bias conditions, while most designers desire the resistance values of the reverse-biased gate. Although the bias conditions of the measurement do not match the typical bias conditions of operation, these measurements at least give good estimates of the desired resistance values—particularly for the R_s and R_d resistances.

Other methods of extracting parasitic resistance values have been presented [3–7]; however, they involve optimization to measured S-parameters, and the optimized values may be sensitive to the initial value choices, as well as to the optimization technique used [8]. None of these methods, dc or S-parameter, directly determines the resistances, thus ascertaining the most accurate approach is difficult. Differences between the resistances derived using the various methods may be on the order of 25%.

While the Fukui method [9] is well suited to the characterization of MESFETs, it will produce inaccurate results with HEMTs. In the reverse-biased HEMT, the electron flow is primarily confined within the 2-DEG, and should have a low resistance. However, in the forward-biased condition, dc current flows through the AlGaAs layer as well, resulting in parasitic resistance values that are inaccurate. The Fukui technique applied to HEMTs can produce estimates of R_s and R_d that appear to be reasonable approximations of the desired resistance levels. Use of this method to determine R_g, however, typically yields a resistance value much larger than desired. The results of using this method with a 0.3×200-μm HEMT are demonstrated in Figures 4.6(a) and 4.6(b). In addition to R_g being too large, all of the parasitic resistances are very sensitive to the applied gate voltage, making it unclear as to which values should be used. Determination of the parasitic resistances in HEMTs usually requires the use of measured S-parameters (as discussed in Section 4.2) or specialized test structures.

One example of specialized test structures that can be used to determine something about the parasitic resistance values is an array of HEMT devices with

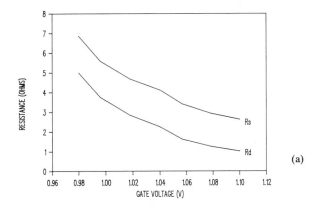

Figure 4.6 Parasitic resistances of a 0.3×200-μm HEMT extracted using forward gate bias techniques. The large bias dependance illustrates the failure of this method as applied to HEMT structures: (a) drain and source resistances; (b) gate resistance. Note that the extracted gate resistance value is unrealistically large.

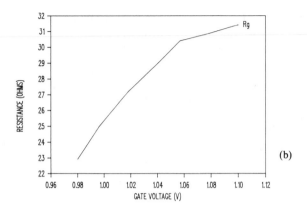

Figure 4.6 continued

varied gate-source or gate-drain spacings. The total drain-source resistance of each device is measured by applying a small drain-source bias (approximately 50 mV) and measuring the induced current. The ratio of these values gives the total resistance:

$$R_{ds} = R_d + R_{ch} + R_s \tag{4.5}$$

where R_{ch} represents the resistance of the channel underneath the gate. The resulting total resistance is then plotted as a function of the terminal spacing. Extrapolation of the data to a terminal spacing of zero produces an estimate for R_{ch}. The sum of $R_d + R_s$ can then be computed using equation (4.5). This method does not provide information concerning the individual R_d and R_s values and does not address R_g at all. Such information, however, can often be used to advantage during parameter extraction.

4.1.3   Rectifying Contact Properties

The dc measurements used to determine R_d, R_g, and R_s for MESFETs also facilitate the characterization of the nonlinear properties of the gate junction when it is driven into the forward-biased state. This is true for HEMT as well as MESFET devices. As described previously, the Schottky junction may be represented by a pair of diodes, one from the gate to the source, and the other from the gate to the drain terminals. The ideality constant n and the reverse saturation current I_s determine the conduction properties of these diodes when the gate is forward biased.

The ideality factor may be determined by selecting two points on the linear portion of the I_g-V_g curve and using the voltage and current values in equation (4.6):

$$n = \frac{q(V_2 - V_1)}{kT \ln(I_2/I_1)} \tag{4.6}$$

where I_1 and I_2 are two measured currents and V_1 and V_2 are the corresponding voltage levels.

Once the ideality constant has been determined, the reverse saturation current may be found using equation (4.7), where I_g is a value of current on the linear portion of the curve and V_g is the corresponding gate voltage:

$$I_s = \frac{I_g}{[e^{qV_g/nkT)} - 1]} \tag{4.7}$$

4.1.4   Gate-Drain Breakdown

The power-handling capabilities of FETs and HEMTs are limited primarily by the gate-drain breakdown voltage. This breakdown is the result of the large electric field produced at the drain-edge of the gate when the device is biased well into saturation [10].

The dc breakdown voltage of a device may be determined by measuring the dc current through the drain as V_{ds} is increased. When the potential across the gate-drain interface reaches the breakdown voltage, current will flow through the reverse-biased gate. This behavior is illustrated in Figure 4.7.

For gate-drain potentials greater than the breakdown voltage, the drain current increases rapidly. The breakdown phenomenon is complex. Frensley [10], however, has derived a semiempirical expression for estimating the breakdown voltage based on the channel doping density (N_d) and epilayer thickness (a). For a homogeneously doped channel, this voltage is:

$$V_{br} = \frac{4.4 \times 10^{13}}{N_d a} \tag{4.8}$$

This equation often gives a good estimate of the breakdown voltage. The dc breakdown characteristics, however, do not accurately reflect the RF breakdown behavior. For dc current, breakdown occurs rapidly as the potential is increased above the breakdown voltage. Pulsed I-V characteristics, however, have been used to show that breakdown for RF signals occurs more gradually [11]. As a result of this difference, the use of dc breakdown voltages to predict RF power-handling capabilities will yield values lower than those actually attainable.

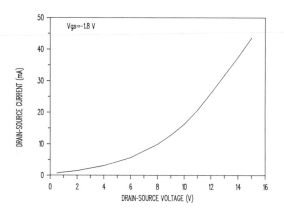

Figure 4.7 Drain-source current as a function of the drain-source bias of a 0.5×300-μm MESFET. Gate-source voltage is -1.8 V.

4.2 EXTRACTION OF SMALL-SIGNAL EQUIVALENT CIRCUITS

Using the techniques described in this section, a small-signal equivalent circuit for any set of measured device S-parameters can be extracted. From these equivalent circuits, the bias-dependent RF element values of output conductance $g_{ds}(V_{ds}, V_{gs})$, transconductance $g_m(V_{ds}, V_{gs})$, gate-source capacitance $C_{gs}(V_{ds}, V_{gs})$, and gate-drain capacitance $C_{gd}(V_{ds}, V_{gs})$ can be obtained. The other intrinsic element values, C_{ds}, R_i, and τ, are also extracted in this process, although their bias dependence is much weaker than the other four elements. Typically, these element values are held constant at their $I_{dss}/2$ value or some other bias in the saturation region. At higher frequencies, the effects of parasitic bond wire inductance or metalization inductance (MMIC devices) become evident, requiring that these effects be included in the small-signal model. This section describes techniques that can be used to extract the intrinsic element values as well as the parasitic inductance values from each set of S-parameters. In addition, Section 4.2.4 describes a technique for extracting the parasitic resistances from the S-parameters (as opposed to the technique described in Section 4.1.2). This technique is very useful when performing parameter extractions for HEMTs.

4.2.1 Conventional Optimization Techniques

Consider the MESFET equivalent circuit shown in Figure 4.8. A commercial simulator such as Super Compact, HP MDS, or Touchstone can be utilized easily to

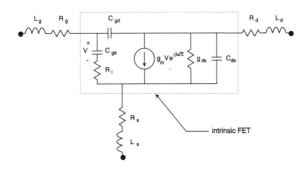

Figure 4.8 MESFET/HEMT small-signal model including parasitic elements.

optimize the 13 elements so as to obtain the best match between the measured and modeled S-parameters. However, unique values of many of the circuit elements are difficult to obtain because the final optimized values depend on the starting values [12].

When small-signal characterization is performed, a set of peak errors, one for amplitude and one for phase, can be defined for each of the S-parameters [12]. As a result of S-parameter fluctuation due to various sources of measurement errors, a match of the equivalent circuit S-parameters to the measured S-parameters within these peak errors can be considered to be an acceptable solution. The circuit in Figure 4.9 has been used to determine a range of each element value that would produce a solution within the peak error. The elements are perturbed individually while optimizing the remaining ones. The range $P_0 - \Delta P$ to $P_0 + \Delta P$ represents values through which the perturbed element can vary with the optimization error that is less than the peak error. Optimization was performed over the frequency range of 1 to 18 GHz. The range of element values for this example is shown in Table 4.1. The quantities P_0 and ΔP are related to the maximum, P_{max}, and minimum, P_{min}, parameter values through the following relations:

$$P_0 = \frac{P_{max} + P_{min}}{2} \tag{4.9}$$

$$\Delta P = \frac{P_{max} - P_{min}}{2} \tag{4.10}$$

The percentage variation is most serious for R_i, τ, L_g, and L_s. For this minimal equivalent circuit, the intrinsic element values can be determined with little uncertainty. However, the elements R_g, R_d, R_s, and L_d were left out of the optimization because they could not be resolved with any certainty. Clearly, such a technique leaves much to be desired.

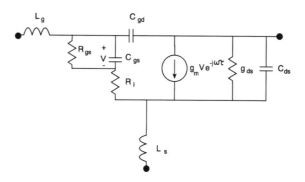

Figure 4.9 Minimal FET equivalent circuit as defined in Reference [12]. Elements shown are those that can be resolved uniquely from optimization within the tolerance specified in Table 4.1

Table 4.1
Range of Extracted Element Values Using Optimization for the Equivalent Circuit in Figure 4.9
(Frequency Range is 1 to 18 GHz [12])

Element Value	P_0	ΔP (%)
R_i (Ω)	11.3	6.75
R_{gs} (kΩ)	1.54	2.12
R_{ds} (Ω)	245	2.14
C_{gs} (fF)	242	1.65
C_{gd} (fF)	35.1	2.25
C_{ds} (fF)	80.2	1.97
g_{mo} (mS)	34.9	0.33
τ (ps)	2.74	10.8
L_g (pH)	21.5	54.8
L_s (pH)	25.4	12.2

In addition to the inability of optimization techniques to resolve uniquely all the equivalent circuit elements in Figure 4.8, another serious drawback to this method is speed. To apply this technique to 25 or so S-parameter sets would be prohibitively time-consuming, literally taking days to accomplish. A much faster and more reliable technique, which has gained widespread acceptance in both the literature and in practice, is the direct extraction method, which is the subject of the following section.

4.2.2 Direct Extraction of Intrinsic Elements

The intrinsic FET topology is such that a y-parameter analysis of the equivalent circuit results in relatively simple expressions that can be equated to measured y-parameter data. Minasian [13] first described such a technique, and since then a number of other authors have used variations of this method to obtain small-signal models for both MESFETs [6, 14] and HEMTs [15]. For this analysis, the equivalent circuit of Figure 4.10 is referenced. Initially, the inductances will be neglected, and discussion of these elements is delayed until Section 4.2.3. In addition, we assume that parasitic resistances have been determined using one of the techniques described in Section 4.1.2 or some other technique. The first step in the extraction process is the conversion of the measured S-parameters to z-parameters using the following:

$$Z'_{11\text{meas}} = \frac{(1 + S_{11})(1 - S_{22}) + S_{12}S_{21}}{\Delta} \tag{4.11}$$

$$Z'_{12\text{meas}} = \frac{2S_{12}}{\Delta} \tag{4.12}$$

$$Z'_{21\text{meas}} = \frac{2S_{21}}{\Delta} \tag{4.13}$$

$$Z'_{22\text{meas}} = \frac{(1 - S_{11})(1 + S_{22}) + S_{12}S_{21}}{\Delta} \tag{4.14}$$

where

$$\Delta = (1 - S_{11})(1 - S_{22}) - S_{12}S_{21} \tag{4.15}$$

and the prime denotes normalized (to Z_0) z-parameters. From the denormalized extrinsic z-parameters, the parasitic resistance values can be de-embedded to obtain the intrinsic z-parameters:

$$z_{11\text{meas}} = Z_{11\text{meas}} - (R_g + R_s) \tag{4.16}$$

$$z_{12\text{meas}} = Z_{12\text{meas}} - R_s \tag{4.17}$$

$$z_{21\text{meas}} = Z_{21\text{meas}} - R_s \tag{4.18}$$

$$z_{22\text{meas}} = Z_{22\text{meas}} - (R_d + R_s) \tag{4.19}$$

where the known extrinsic values are denoted by uppercase letters and the intrinsic values are denoted by lowercase letters. We now convert these z-parameter data into y-parameter data:

$$y_{11meas} = z_{22meas}/(|z|Z_0) \tag{4.20}$$

$$y_{12meas} = -z_{12meas}/(|z|Z_0) \tag{4.21}$$

$$y_{21meas} = -z_{21meas}/(|z|Z_0) \tag{4.22}$$

$$y_{22meas} = z_{11meas}/(|z|Z_0) \tag{4.23}$$

where

$$|z| = z_{11meas}z_{22meas} - z_{12meas}z_{21meas} \tag{4.24}$$

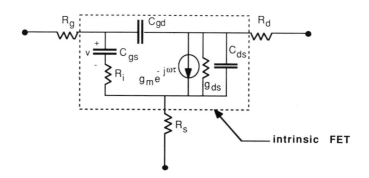

Figure 4.10 MESFET/HEMT small-signal model without parasitic inductances.

Equations (4.20) through (4.24) provide de-embedded y-parameter data from the measured S-parameters. From the equivalent circuit in Figure 4.10, the exact analytical form of the intrinsic y-parameters can be derived:

$$y_{11} = R_iC_{gs}^2\omega^2/D + j\omega(C_{gs}/D + C_{gd}) \tag{4.25}$$

$$y_{12} = -j\omega C_{gd} \tag{4.26}$$

$$y_{21} = \{g_me^{(-j\omega\tau)}/(1 + jR_iC_{gs}\omega)\} - j\omega C_{gd} \tag{4.27}$$

$$y_{22} = g_{ds} + j\omega(C_{ds} + C_{gd}) \tag{4.28}$$

where $D = 1 + \omega^2 C_{gs}^2 R_i^2$. From these four equations, closed-form expressions for extracting the element values from y-parameter data can be developed.

4.2.2.1 Capacitance Extraction

The simplest element to determine is the gate-drain capacitance C_{gd} because it is found directly from Im[$y_{12\text{meas}}$]. An optimum C_{gd} (in the y-domain) can be found over some frequency range from the slope of a linear least-squares fit to the measured data:

$$C_{gd} = -m_{y12} \tag{4.29}$$

where m_{y12} is the slope of the regression line to Im[$y_{12\text{meas}}$] *versus* ω data. Likewise, the sum $(C_{gd} + C_{ds})$ can be computed from the best fit line to the Im[$y_{22\text{meas}}$] data. The drain-source capacitance is then given by

$$C_{ds} = m_{y22} - C_{gd} \tag{4.30}$$

Extraction of C_{gs} is complicated by the presence of the D term in the denominator of equation (4.25). However, under the low frequency assumption $(\omega C_{gs} R_i)^2 \ll 1$, Im[y_{11}] simplifies to $j\omega(C_{gs} + C_{gd})$. The gate-source capacitance can now be extracted from

$$C_{gs} = m_{y11} - C_{gd} \tag{4.31}$$

where m_{y11} is the slope of the regression line to Im[$y_{11\text{meas}}$] data taken over a low frequency range.

   In Figure 4.11, values of Im[$y_{12\text{meas}}$], Im[$y_{22\text{meas}}$], and Im[$y_{11\text{meas}}$] *versus* frequency are shown for a 0.5×300-μm ion-implanted MESFET. The data are very linear through 2.5 GHz, indicating a high correlation data fit. Because of the excellent linearity exhibited by the data from which capacitance is extracted, in most cases, equations (4.29) through (4.31) can often be applied at a single frequency point. For this case, the expressions reduce to

$$C_{gd} = -\text{Im}[\,y_{12}]/\omega \tag{4.32}$$

$$C_{ds} = \text{Im}[\,y_{22}]/\omega - C_{gd} \tag{4.33}$$

$$C_{gs} = \text{Im}[\,y_{11}]/\omega - C_{gd} \tag{4.34}$$

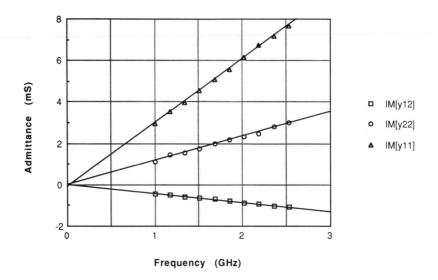

Figure 4.11 Imaginary parts of y_{12}, y_{22}, and y_{11} *versus* frequency for a 0.5×300-μm GaAs MESFET biased at $V_{gs} = -0.8$ V and $V_{ds} = 3.0$ V. Device capacitances are extracted from the slopes of the plots per equations (4.29) through (4.31).

4.2.2.2 Extraction of g_{ds} and g_m

From equation (4.28), g_{ds} is given by

$$g_{ds} = \text{Re}[\, y_{22}] \tag{4.35}$$

Although Re[$y_{22\text{meas}}$] falls off at high frequency, the value of g_{ds} computed from low frequency Re[$y_{22\text{meas}}$] data provides a reliable value for matching the S-parameters over frequency. The data are fairly flat in the low frequency range, so g_{ds} may be computed from either a single point or from the average of a few points.

A simple expression for g_m is not obtainable directly from equation (4.27) as was the case for output conductance. However, a simple expression for g_m, which proves to be an excellent approximation at low frequencies, can be obtained. From equation (4.27), transconductance in complex rectangular form is expressed as

$$g_{mr} + jg_{mi} = g_m e^{(-j\omega\tau)} \tag{4.36}$$

where

$$g_{mr} = \text{Re}[\, y_{21}] - \text{Im}[\, y_{21}]R_i C_{gs}\omega - \omega^2 C_{gd}C_{gs}R_i \tag{4.37}$$

and

$$g_{mi} = \text{Re}[\, y_{21}]R_i C_{gs}\omega + \text{Im}[\, y_{21}] + \omega C_{gd} \tag{4.38}$$

The transconductance is simply the magnitude of these two components:

$$g_m = (g_{mr}^2 + g_{mi}^2)^{1/2} \tag{4.39}$$

for a typical low noise device at low frequencies (≈ 2 GHz) biased around $I_{dss}/2$, $\text{Re}[\, y_{21}] \approx 45$ mS, $\text{Im}[\, y_{21}] \ll \text{Re}[\, y_{21}]$, $R_i C_{gs}\omega \approx 0.01$, and $\omega C_{gd} \approx 0.001$. Therefore, $\text{Re}[\, y_{21}]$ is the only term of significance and g_m is well approximated as

$$g_m \approx \text{Re}[\, y_{21}]. \tag{4.40}$$

As with output conductance, the $\text{Re}[\, y_{21\text{meas}}]$ data are smooth, and g_m can be computed from either a single point or from an average of a few points.

4.2.2.3 Extraction of τ and R_i

The parameters τ and R_i are by far the most difficult of the equivalent circuit elements shown in Figure 4.10 to extract from the measured data. The primary reason for this is that the data from which these values are derived are often quite noisy. This is especially true of data taken at low frequencies. Unfortunately, the low frequency range is where the simple expressions are valid and, consequently, is where the extraction process is most easily performed. Shown in Figure 4.12 is $\text{Re}[\, y_{11\text{meas}}]$

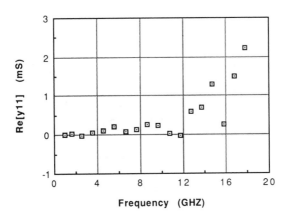

Figure 4.12 Real part of y_{11} *versus* frequency for a 0.5×300-μm GaAs MESFET biased at $V_{gs} = -0.8$ V and $V_{ds} = 3.0$ V. Charging resistance R_i is extracted from these data.

versus frequency. Note the large percentage variation in Re[$y_{11\text{meas}}$], especially at low frequency. Extraction of reliable values for R_i is aided if high frequency data rather than low frequency data are considered. This also helps in the extraction of τ because, from equation (4.27) any expression for τ will include an R_i dependence. In addition, because τ is primarily a parameter that affects the high frequency phase of S_{21}, it is logical to extract it in this frequency range.

Equating real parts in equation (4.25) and solving for R_i, a quadratic equation in R_i is obtained. If we solve for R_i, we obtain

$$R_i = \frac{1 - \left(1 - \dfrac{4\text{Re}^2[\,y_{11}]}{\omega^2 C_{gs}^2}\right)^{1/2}}{2\text{Re}[\,y_{11}]} \tag{4.41}$$

where C_{gs} is given by evaluation of equation (4.34). Equation (4.41) can be applied to one or more data points at the high end of the measured data spectrum; the values are then averaged to obtain acceptable results. Figure 4.13 shows the values of R_i computed from equation (4.41) over frequency for a low noise MESFET. The data scatter is significant at low frequencies and diminishes at higher frequencies. The noise inherent to the Re[$y_{11\text{meas}}$] data is very process and device dependent. Significant scatter seems to be present in all the MESFETs, but is not present to any significant degree in most HEMTs. Obviously, when confronted with noisy values of Re[$y_{11\text{meas}}$], an uncertainty is inherent to R_i extraction. However, use of equation (4.41) at high frequencies seems to minimize this uncertainty while main-

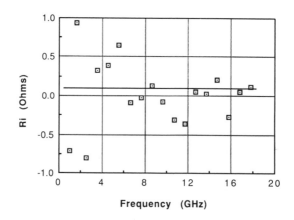

Figure 4.13 Extracted value of R_i over frequency for a 0.5×300-μm GaAs MESFET biased at $V_{gs} =$ 0.8 V and $V_{ds} = 3.0$ V. Results are computed from equation (4.41) using the data in Figure 4.12.

taining the simplicity of a closed-form expression. Because R_i primarily affects S_{11}, a more reliable method for extracting it may be to optimize it to match S_{11} after the other elements have been extracted [16]. An important point to be made here is that, although extracting R_i to noisy data can give widely varying values and consequently a less than perfect match to S_{11}, the overall effect of R_i on the resulting S-parameters of the circuit model is second order at best. This is especially true considering that the element R_i provides no unique equivalent circuit performance capabilities and is of questionable physical significance. The primary effects of R_i on the equivalent circuit behavior are to increase the real portion of the modeled device input impedance and the time required for the channel current to respond to gate voltage fluctuations. Note, however, that the resistor R_g also increases the real part of the modeled input impedance, and the element τ increases the channel current to gate voltage response time.

The element τ is important to the accurate modeling of observed high frequency phase characteristics in S_{21}. At low frequencies, the approximations $\omega\tau \ll 1$ and $D = 1$ are good, and we obtain a simple expression for τ:

$$\tau = (-\text{Im}[\,y_{21}]/\omega - g_m R_i C_{gs} - C_{gd})/g_m \tag{4.42}$$

Previous authors [6, 13, 14] have utilized this low frequency approximation to obtain τ from low frequency data. However, τ is primarily a parameter that affects high frequency device response. At high frequencies, the approximations used in obtaining equation (4.42) begin to break down. This is especially true for HEMTs that can be used at frequencies well above 26 GHz. At this frequency, for a value $\tau = 2.0$ ps, $\omega\tau = 0.33$, and the approximation $\omega\tau \ll 1$ is no longer valid. An exact expression for τ from equations (4.37) and (4.38) can be developed:

$$\tau = -\frac{1}{\omega}\tan^{-1}\left(\frac{g_{mi}}{g_{mr}}\right) \tag{4.43}$$

which is valid over all frequencies. Hence, (4.43) can be used to compute τ at the highest frequency for which accurate data are available. Note that the value computed for the time constant expressed by equation (4.43) is somewhat dependent on the choice of source inductance, L_S. If the value for source inductance is poorly chosen, then the time constant value will be highly frequency dependent. Shown in Figure 4.14 is the value of τ computed from equations (4.42) and (4.45) *versus* frequency. Note the slight frequency dependence of both expressions and the deviation at high frequency between the τ values computed by (4.42) and (4.43). The important point to be made here is that whichever expression is used, the expression should be applied at frequencies well above those used to evaluate the other element values.

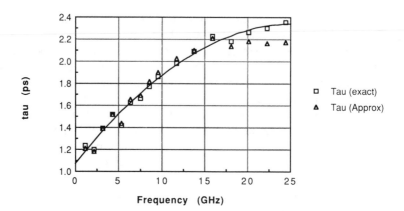

Figure 4.14 Extracted value of τ *versus* frequency for a $0.5 \times 300\text{-}\mu\text{m}$ GaAs MESFET biased at $V_{gs} = -0.8$ V and $V_{ds} = 3.0$ V. Values were extracted from both the approximate (low frequency) expression (4.42) and the exact expression (4.43).

In summary, the expressions for extracting the equivalent circuit elements of Figure 4.10 can be divided into two groups. For those extracted at frequencies where $(\omega C_{gs} R_i)^2 \ll 1$:

$$C_{gd} = -\text{Im}[\, y_{12}]/\omega \tag{4.44}$$

$$C_{ds} = \text{Im}[\, y_{22}]/\omega - C_{gd} \tag{4.45}$$

$$C_{gs} = \text{Im}[\, y_{11}]/\omega - C_{gd} \tag{4.46}$$

$$g_{ds} = \text{Re}[\, y_{22}] \tag{4.47}$$

$$g_m = \text{Re}[\, y_{21}] \tag{4.48}$$

and for those extracted at high frequencies:

$$R_i = \frac{1 - \left(1 - \dfrac{4\text{Re}^2[\, y_{11}]}{\omega^2 C_{gs}^2}\right)^{1/2}}{2\text{Re}[\, y_{11}]} \tag{4.49}$$

$$\tau = -\frac{1}{\omega} \tan^{-1}\left(\frac{g_{mi}}{g_{mr}}\right) \tag{4.50}$$

where

$$g_{mr} = \text{Re}[\, y_{21}] - \text{Im}[\, y_{21}]R_i C_{gs}\omega - \omega^2 C_{gd} C_{gs} R_i \tag{4.51}$$

$$g_{mi} = \text{Re}[\, y_{21}]R_i C_{gs}\omega + \text{Im}[\, y_{21}] + \omega C_{gd} \tag{4.52}$$

4.2.2.4 Comparison of Measured and Modeled S-Parameter Data

Although the small-signal equivalent circuit elements are extracted from y-parameter data, the objective is to match measured S-parameters as closely as possible to an equivalent circuit. To evaluate the quality of the fit to measured data, definition of an error term is convenient:

$$E_{ij} = \frac{1}{n} \sum_{k=1}^{n} \frac{|S^k_{ij\text{meas}} - S^k_{ij\text{mod}}|}{|S^k_{ij\text{meas}}|} \qquad (4.53)$$

where n is the number of discrete frequencies used.

Figure 4.15 shows the measured *versus* modeled S-parameters of an ion-implanted GaAs MESFET. The element values were extracted using the expressions in this Section and are given in the first column of Table 4.2. The given errors are expressed as a percentage of equation (4.53). The model demonstrates an excellent match to the measured data at low frequency. The deviation between measured

□ S11 MEASURED

+ S22 MEASURED

◇ S11 MODELED

× S22 MODELED

f1: 1. 00000
f2: 18. 0000

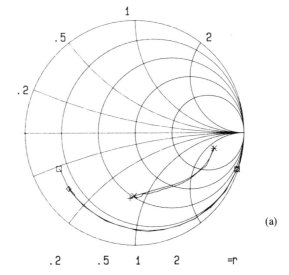

(a)

Figure 4.15 Measured *versus* modeled S-parameters for a 0.5×300-μm GaAs MESFET biased at V_{gs} = −0.8 V and V_{ds} = 3.0 V. Inductances were assumed to be zero. Equivalent circuit element values are given in column 1 of Table 4.2. Frequency range is 1 to 18 GHz: (a) S_{11} and S_{22}; (b) S_{12}; (c) S_{21}.

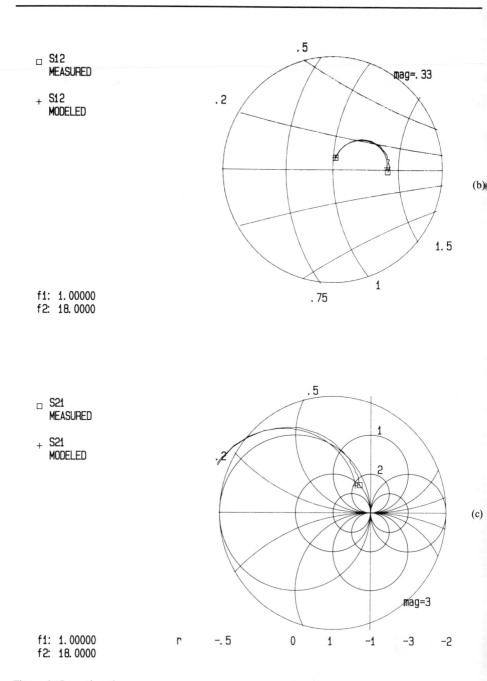

□ S12
MEASURED

+ S12
MODELED

.5

mag=. 33

.2

(b)

1.5

1

.75

f1: 1.00000
f2: 18.0000

□ S21
MEASURED

+ S21
MODELED

.5

1

.2

2

(c)

mag=3

f1: 1.00000
f2: 18.0000

r   -.5   0   1   -1   -3   -2

Figure 4.15 continued

Table 4.2
Extracted Element Values for a MMIC GaAs FET and HEMT Using the Algorithm of Section 4.2.3*

Quantity	GaAs FET		HEMT	
	First Iteration	Convergence	First Iteration	Convergence
C_{gd} (fF)	65.67	65.41	35.32	31.41
C_{gs} (fF)	398.3	396.5	561.2	504.3
C_{ds} (fF)	109.2	110.0	49.36	51.39
r_{ds} (Ω)	292.9	291.0	1357	937.1
g_m (mS)	46.38	46.20	67.63	62.52
τ (ps)	2.76	2.13	2.86	2.32
R_i (Ω)	0.0	0.0	0.160	0.434
L_s (pH)	0.0	3.4	0.0	2.0
L_d (pH)	0.0	0.0	0.0	91.0
L_g (pH)	0.0	62.0	0.0	89.0
E_{11} (%)	11.4	0.84	20.7	1.46
E_{12} (%)	4.1	2.21	11.1	2.31
E_{21} (%)	5.6	3.62	11.8	2.12
E_{22} (%)	4.0	2.38	7.57	1.74
E_{tot} (%)	6.3	2.3	12.8	1.91

*Values are given for inductances set to zero (columns 1 and 3) and after convergence of the algorithm (columns 2 and 4). Parasitic resistances are: GaAs FET, R_g = 2.0 Ω, R_s = 3.1 Ω, and R_d = 3.6 Ω; HEMT, R_g = 6.5 Ω, R_s = 2.35 Ω, and R_d = 4.93 Ω. Frequency range is 1 to 18 GHz (GaAs FET), 1 to 26 GHz (HEMT). Input and output pad capacitance (20 fF) was de-embedded from HEMT data prior to extraction.

and modeled values at higher frequencies is primarily due to the absence of inductances in the model. In Section 4.2.3, the methods of Section 4.2.2 are extended to compute parasitic inductances as well as intrinsic device equivalent circuit element values.

4.2.3 Parasitic Inductance Extraction

As demonstrated by the data in Figure 4.15, the inclusion of parasitic inductances in the model is critical to accurate modeling of MESFET/HEMT high frequency RF behavior. A number of authors have addressed the problem of inductance extraction for MESFETs [3, 6]. The principal technique used is the cold-chip technique proposed by Diamand and Laviron [3]. The drawback to this method is that it requires measurement of the device S-parameters at 0 V drain-source bias. This presents an extra step in the characterization process.

In addition, although this technique has proved useful in extracting extrinsic package inductance, whether the cold-chip technique is valid for MMIC devices is not clear. In MMIC devices, the parasitic inductances result from metalization strips at the gate and drain terminals of the device. This metalization inductance is a function of the substrate (semiconductor) properties, which are significantly different under a zero- or forward-bias condition than they are under reverse bias. Ladbrooke [17] gives the following expression for the inductance of a gate strip in a MMIC GaAs FET:

$$L_g = \mu_0 d_{ch} Z / m^2 L \tag{4.54}$$

where Z is the gate width, d_{ch} is the depletion depth under the gate, m is the number of gate fingers, L is the gate length, and μ_0 is the permeability of free space. Based on this expression, we have good reason to believe that the value for L_g extracted under zero or forward bias can be significantly different from the value under normal (reverse-bias) operation. The method described in Reference [14] and outlined in this section obtains the parasitic inductance from the same s-parameter measurements from which the intrinsic elements are extracted. The method is fast and equally valid for either MMIC or packaged devices.

Referring to Figure 4.8, the extrinsic device z-parameters can be expressed in terms of the intrinsic parameters as follows:

$$Z_{11} = z_{11} + (R_g + R_s) + j\omega(L_g + L_s) \tag{4.55}$$

$$Z_{12} = z_{12} + R_s + j\omega L_s \tag{4.56}$$

$$Z_{21} = z_{21} + R_s + j\omega L_s \tag{4.57}$$

$$Z_{22} = z_{22} + (R_d + R_s) + j\omega(L_d + L_s) \tag{4.58}$$

where the extrinsic device z-parameters are denoted by uppercase letters. The key to this extraction technique lies in the fact that at sufficiently low frequencies, the inductances have a minimal effect on the response of the small-signal model. This is evidenced in Figure 4.15, where the measured and modeled data match at the lower frequencies. Because the intrinsic elements are extracted at low frequencies, the assumption is valid that intrinsic elements extracted in the absence of parasitic inductance are close to the actual values for the complete model. From equations (4.55) through (4.58), the inductances can be computed at *high* frequencies by equating imaginary components of the measured and modeled data:

$$\text{Im}[Z_{11\text{meas}}] - \text{Im}[z_{11\text{mod}}] = \Delta Z_{11} = \omega(L_g + L_s) \tag{4.59}$$

$$\text{Im}[Z_{12\text{meas}}] - \text{Im}[z_{12\text{mod}}] = \Delta Z_{12} = \omega L_s \tag{4.60}$$

$$\text{Im}[Z_{21\text{meas}}] - \text{Im}[z_{21\text{mod}}] = \Delta Z_{21} = \omega L_s \tag{4.61}$$

$$\text{Im}[Z_{22\text{meas}}] - \text{Im}[z_{22\text{mod}}] = \Delta Z_{22} = \omega(L_d + L_s) \tag{4.62}$$

The inductance values are then easily obtained from the ΔZ_{ij} data as follows:

(1) L_s from ΔZ_{12} or ΔZ_{21}.
(2) L_g from ΔZ_{11}.
(3) L_d from ΔZ_{22}.

Figure 4.16 plots equation (4.59) for a MMIC GaAs FET. The slope of the line represents the inductance $(L_s + L_g)$. The source inductance for this particular device was approximately zero, so the gate inductance can be obtained directly from this plot and is ≈ 60 pH.

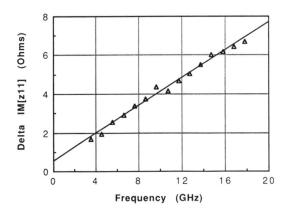

Figure 4.16 Plot of ΔZ_{11} *versus* frequency for a 0.5×300-μm GaAs MESFET biased at $V_{gs} = -0.8$ V and $V_{ds} = 3.0$ V. The sum $(L_g + L_s)$ is obtained from the slope of the plot.

The value of L_s extracted can vary considerably depending on whether we use ΔZ_{12} or ΔZ_{21} to find it. For the devices looked at here, ΔZ_{12} was found to give the best results. Because the determination of L_s by either method requires relatively simple calculations, in many cases, performing the calculation both ways and then determining which provides the least error may be helpful. Alternatively, a weighted average can be used to compute L_s from

$$L_s = \frac{(W_{12}L_s^{12} + W_{21}L_s^{21})}{(W_{12} + W_{21})} \tag{4.63}$$

where W_{12} and W_{21} are the weighting factors and L_s^{12} and L_s^{21} are the inductances extracted from ΔZ_{12} or ΔZ_{21}, respectively. The weighting factors can either be selected by trial and error or varied automatically in the extraction algorithm to obtain the desired trade-off between the modeling errors in Z_{12} and Z_{21}.

Successively better approximations to both the intrinsic and extrinsic element values can be obtained by repeating this extraction procedure, but de-embedding the inductance found in the previous iteration in all subsequent iterations. Diminishing returns in increased accuracy are usually achieved after two or three iterations. The algorithm for the extraction of intrinsic elements and parasitic inductors is as follows:

(1) Convert measured S-parameters to z-parameters using equations (4.11) through (4.15).
(2) Evaluate inductances from equations (4.59) through (4.62) (inductors set to 0 for first iteration).
(3) De-embbed parasitics to obtain low frequency intrinsic z-parameters using equations (4.55) through (4.58).
(4) Convert the intrinsic z-matrix to a y-matrix using equations (4.20) through (4.24).
(5) Evaluate intrinsic elements from equations (4.44) through (4.52).
(6) Compute intrinsic modeled z-parameters from circuit elements found in step 5.
(7) Compute extrinsic modeled z-parameters from equations (4.55) through (4.58) using known parasitic resistance values and inductances found in step 2.
(8) Convert extrinsic modeled z-parameters back to s-parameters.
(9) Compute modeling error, E_{mod} from equation (4.53).
(10) If $E_{mod} < E_{tol}$ then go to 11, otherwise go to 2.
(11) Stop.

The results obtained by application of this algorithm to a MMIC GaAs FET and HEMT device are shown in Table 4.2. The parasitic resistances for the GaAs FET were obtained using dc measurements. The HEMT resistances were obtained using a technique to be outlined in the following subsection. The column 1 and 3 results represent the values obtained under the assumption that the inductors were all zero. The column 2 and 4 results were obtained after three and five iterations of the algorithm, respectively. Note the large decrease in modeling error resulting from the inclusion of parasitic inductances. The modeled S-parameters for these two devices are plotted along with the measured values in Figures 4.17 and 4.18.

4.2.4 Extraction of Parasitic Resistances from S-Parameters

Because of difficulties in extracting HEMT parasitic resistances from dc measurements, the ability to extract these elements from the same RF measurements used to extract the other elements is desirable. Although a unique set of resistance values

□ S11
MEASURED

+ S22
MEASURED

◇ S11
MODELED

× S22
MODELED

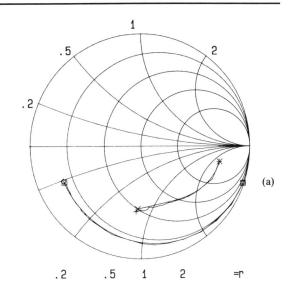

(a)

f1: 1.00000
f2: 18.0000

□ S12
MEASURED

+ S12
MODELED

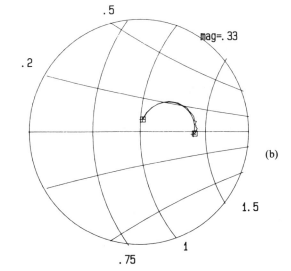

(b)

f1: 1.00000
f2: 18.0000

Figure 4.17 Measured *versus* modeled *S*-parameters for a 0.5×300-μm GaAs MESFET biased at V_{gs} = -0.8 V and V_{ds} = 3.0 V. Equivalent circuit element values are given in column 2 of Table 4.2. Frequency range is 1 to 18 GHz: (a) S_{11} and S_{22}; (b) S_{12}; (c) S_{21}.

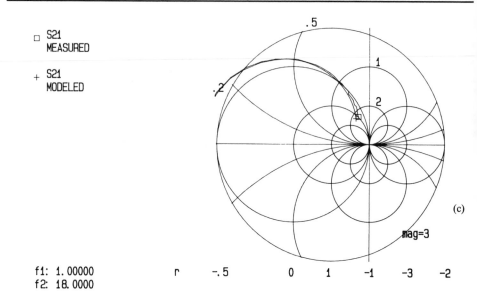

□ S21
  MEASURED

+ S21
  MODELED

f1: 1. 00000
f2: 18. 0000

Figure 4.17 continued

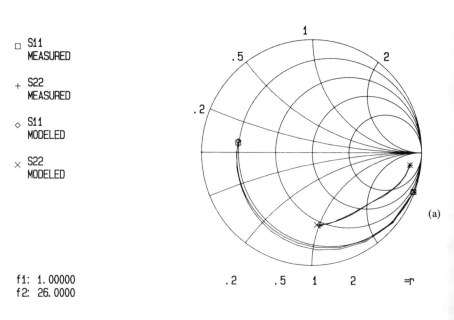

□ S11
  MEASURED

+ S22
  MEASURED

◇ S11
  MODELED

× S22
  MODELED

f1: 1. 00000
f2: 26. 0000

Figure 4.18 Measured *versus* modeled *S*-parameters for a 0.7×200-μm HEMT biased at $V_{gs} = -0.25$ V and $V_{ds} = 3.0$ V. Equivalent circuit element values are given in column 4 of Table 4.2. Frequency range is 1 to 26 GHz: (a) S_{11} and S_{22}; (b) S_{12}; (c) S_{21}.

S12
MEASURED

+ S12
MODELED

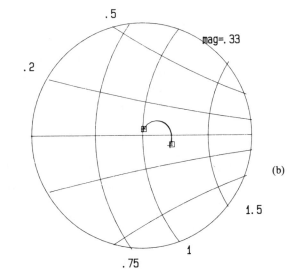

(b)

f1: 1.00000
f2: 26.0000

S21
MEASURED

+ S21
MODELED

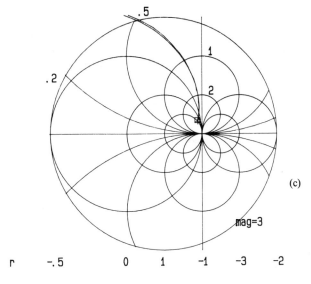

(c)

f1: 1.00000
f2: 26.0000

Figure 4.18 continued

cannot be found directly as with the other elements, a simple optimization algorithm can be used in conjunction with the direct extraction expressions to obtain a set of resistances that minimizes the error defined by (4.53). For a given set of parasitic resistance values R_g, R_s, R_d, and S-parameters, a set of intrinsic element values and parasitic inductance values can be obtained using the techniques already described. Hence, the algorithm need only vary the resistance values by some random or gradient technique until the errors given by (4.53) are minimized.

As mentioned previously, a similarity exists between R_g and R_i in that they both affect the real part of S_{11}. Because of this, both of these elements cannot be extracted uniquely. The technique described here requires that R_g be constrained. The value used for R_g can be obtained from dc measurements or from physical considerations (Section 2.1.1). This leaves R_d and R_s to be optimized. The optimization method used here was the *downhill simplex* method. This is a slow technique, but it proved to be robust. For a given value of R_g, the algorithm always converged to values of R_d and R_s, which minimized the overall S-parameter error. More sophisticated optimization techniques such as those described in Reference [18] should give the same results with greater computational efficiency.

Shown in Table 4.3 are the results of four HEMT extractions; each uses a different value for R_g. The values of R_d and R_s were optimized while the remaining elements were computed directly each time the resistances were updated. Not surprisingly, the total error is relatively insensitive to R_g because the value extracted for R_i changes to account for the variations in R_g. Inspection of the data in Table 4.3 shows the sum of R_g and R_i to be approximately constant, varying from 7 to 8 Ω. Although the total error is approximately the same for each of the four cases, many of the element values vary significantly. This is further demonstration of the fact that no unique equivalent circuit exists for a set of measured S-parameters. A unique equivalent circuit for a given set of parasitic resistances can, however, be extracted. For this example, the column 1 data may be selected because they provide both a low error and a more even distribution of error between the individual S-parameters.

4.2.5 Element Extraction at Multiple Bias Points

An important reason for developing efficient and accurate methods for extracting small-signal models is to obtain the RF bias dependence of the element values for large-signal model extraction. Because most large-signal model topologies only consider $g_{ds}(V_{ds},V_{gs})$, $g_m(V_{ds},V_{gs})$, $C_{gs}(V_{ds},V_{gs})$, and $C_{gd}(V_{ds},V_{gs})$ to be bias dependent, the remaining elements τ, C_{ds}, and the parasitics must be constrained to a constant value in the large-signal analysis. The charging resistance R_i is often neglected altogether in the large-signal analysis. By constraining these element values, some error

Table 4.3

Extracted Element Values for a HEMT Using the Algorithm of Section 4.2.3 Coupled with an Optimization Technique to Obtain R_d and R_s*

Quantity	Element Values			
	$R_g = 6.5\ \Omega$	$R_g = 4.0\ \Omega$	$R_g = 3.0\ \Omega$	$R_g = 1.0\ \Omega$
C_{gd} (fF)	31.41	31.88	32.20	32.36
C_{gs} (fF)	504.2	472.0	450.4	428.5
C_{ds} (fF)	51.39	53.36	53.00	56.47
r_{ds} (Ω)	937.1	980.7	1019	1039
g_m (mS)	62.52	58.98	56.37	55.19
τ (ps)	2.33	2.28	2.32	2.19
R_i (Ω)	0.434	3.75	4.95	6.78
L_s (pH)	2.0	0.0	2.0	1.0
L_d (pH)	91.0	78.0	80.0	54.0
L_g (pH)	89.0	90.0	92.0	95.0
R_d (Ω)	4.93	6.69	6.79	11.2
R_s (Ω)	2.35	1.47	0.81	0.48
E_{11} (%)	1.46	1.17	1.51	2.59
E_{12} (%)	2.31	2.18	1.90	2.52
E_{21} (%)	2.12	2.62	2.67	3.11
E_{22} (%)	1.74	1.95	1.92	1.17
E_{tot} (%)	1.91	1.98	2.00	2.35

*Frequency range is 1 to 26 GHz. Input and output pad capacitance (20 fF) was de-embedded from HEMT data prior to extraction.

is introduced, but overall the match to the measured S-parameters is good. A logical bias place to extract the constrained values is at the operating point of the device in question. In most cases, the values of τ, C_{ds}, and the parasitics do not vary significantly in the saturation region. Hence, if the operating point is not known, the constant element extraction point is somewhat arbitrary.

Shown in Figures 4.19(a) through 4.19(e) are the I-V characteristics and small-signal element bias dependencies for a HEMT. The element values were extracted at 25 separate bias points (the $V_{gs} = -0.8$ V values are not plotted). The constant element values are the values in column 1 of Table 4.3. This entire extraction process took less than two minutes to execute on an IBM-AT machine. The modeling error computed from equation (4.53) is shown in Figure 4.19(f). The error was computed over the frequency range of 1 to 26 GHz. The results show that a good fit to the measured S-parameters can be obtained over bias by using a relatively simple four-element bias-dependent model.

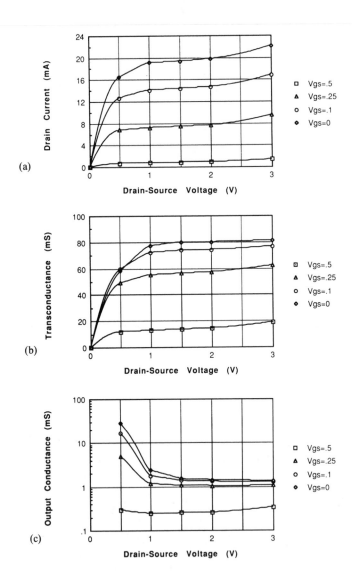

Figure 4.19 Small-signal element values for a 0.7×200-μm HEMT over bias. Constant elements were constrained to $\tau = 2.32$ ps, $C_{ds} = 51.4$ fF, $R_g = 6.5$ Ω, $R_s = 2.35$ Ω, $R_d = 4.93$ Ω, $L_g =$ 89 pH, $L_s = 2$ pH, $L_g = 9$ pH: (a) measured bias points; (b) transconductance; (c) output conductance; (d) gate-source capacitance; (e) gate-drain capacitance; (f) measured *versus* modeled error computed from equation (4.53) as $(E_{11} + E_{12} + E_{21} + E_{22})/4$.

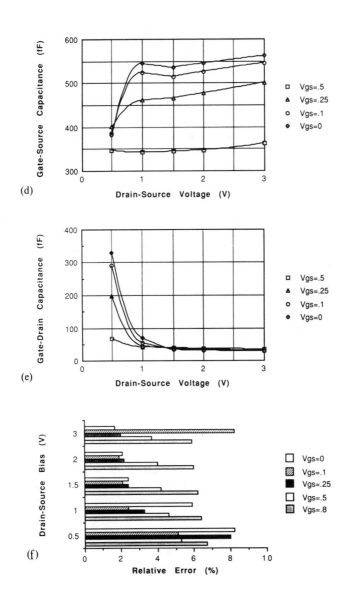

Figure 4.19 continued

4.3 LARGE-SIGNAL EXTRACTION

The empirical and physical parameters contained in large-signal device models are rarely known when devices are fabricated or acquired by the circuit or device designer. These parameters define the electrical characteristics unique to an individual device and must be extracted for use in a large-signal circuit simulation. The parameter values are defined by fitting measured data to modeled data. The measured data can include measured physical parameters such as device geometry or doping density, or electrical parameters such as *I-V* characteristics or *S*-parameters. The problem of which measurements to make in order to characterize fully a device in terms of large-signal behavior is significant. Currently, small-signal *S*-parameters over a varied bias range and harmonic content in the output signal have been used with good success [19, 20].

Some parameters required for a large-signal MESFET or HEMT model can be determined from direct analytical and graphical techniques. Section 4.1 presents methods to obtain parasitic resistance values in MESFETs and Schottky diode parameters for both MESFETs and HEMTs using such techniques. Extraction of most other required parameters, however, is significantly more difficult. The process of determining these other parameters is greatly simplified by using nonlinear optimization techniques. Such techniques are the subject of Sections 4.3.1 through 4.3.4. The development of multidimensional optimization routines is a task that requires familiarity with many specialized concepts. This section first presents a general introduction of some optimization terminology and methods, then applies these methods to the particular problem of large-signal model parameter extraction.

4.3.1 Nonlinear Unconstrained Optimization

Nonlinear optimization is the process of finding the minimum or maximum of a nonlinear function of one or several independent variables. The function being maximized or minimized is called the *objective function*. To visualize a function of this type, an isometric representation of a two-dimensional surface is provided in Figure 4.20. This surface represents a function of two variables, but the idea is easily extended to a function of *n* variables in *n* space. If the problem is solved by allowing the independent variables to assume any value, then it is an unconstrained problem; otherwise it is a constrained problem and the confining conditions are called *constraints*. For most cases, large-signal parameter extraction is solved using unconstrained optimization. The subject of optimization is not trivial and, therefore, this section will only introduce the reader to the basic concepts of optimization followed by a few reliable methods.

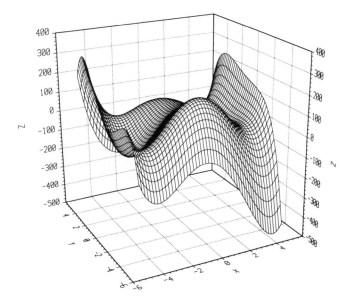

Figure 4.20 Isometric view of a function of two variables.

4.3.2 Fundamentals of Unconstrained Optimization

If the n independent variables of interest are $x_1, x_2, \ldots, x_n$, then the objective function can be written $F(x_1, x_2, \ldots, x_n)$. The vector form of the variables is more convenient:

$$\mathbf{x} = \begin{bmatrix} x_1 \\ x_2 \\ \vdots \\ x_n \end{bmatrix} \tag{4.64}$$

In the case of large-signal parameter extraction, the variables of equation (4.64) are the parameters in the large-signal model equations.

The objective function can now be written as $F(\mathbf{x})$. The gradient vector of $F(\mathbf{x})$ can be written as

$$\mathbf{g}(\mathbf{x}) = \begin{bmatrix} \partial F/\partial x_1 \\ \partial F/\partial x_2 \\ \vdots \\ \partial F/\partial x_n \end{bmatrix} = \nabla F(\mathbf{x}) \tag{4.65}$$

where the del operator is defined as $(\partial/\partial x_1, \ldots, \partial/\partial x_n)^T$. If $F(\mathbf{x})$ is twice continuously differentiable, then a matrix of second partial derivatives exists called the *Hessian matrix:*

$$\mathbf{G}(\mathbf{x}) = \begin{bmatrix} \partial^2 F/\partial x_1^2 & \ldots & \partial^2 F/\partial x_1 \partial x_n \\ \vdots & & \vdots \\ \partial^2 F/\partial x_n \partial x_1 & \ldots & \partial^2 F/\partial x_n^2 \end{bmatrix} \qquad (4.66)$$

which can be expressed as $\mathbf{G}(\mathbf{x}) = \partial^2 F(\mathbf{x})$. Note that for the particular problem of interest, the first and second derivatives specified in equations (4.65) and (4.66) are derivatives of the modeled error F, with respect to the large-signal model parameter vector $\mathbf{x}$. For large-signal parameter extraction, the Hessian matrix is square and symmetric.

These expressions can be used to determine the derivative along any arbitrary line $\mathbf{z}$. Let $\mathbf{e}$ be the unit vector along $\mathbf{z}$ so that $\mathbf{e} = \mathbf{z}/\|\mathbf{z}\|_2$. The directed first derivative (slope) of $F(\mathbf{x})$ along $\mathbf{z}$ is

$$\mathbf{e}^T \mathbf{g}(\mathbf{x}) \qquad (4.67)$$

and the corresponding second derivative (curvature) is

$$\mathbf{e}^T \mathbf{G}(\mathbf{x}) \mathbf{e} \qquad (4.68)$$

The ultimate goal of optimization is to find the global minimum of the objective function. In terms of large-signal parameter extraction problems, the error function defining the difference between the measured and modeled data is minimized. The set of parameters at the minimum is called the *minimizer*. The minimizer of a function can take on several forms and is usually expressed as x^*. A *strong minimum* is a point at which the objective function increases locally in all directions. A *weak minimum* is a point at which the objective function remains locally the same in some directions but increases locally in others. A function may have any number of these points all of which are *local* minimizers. A *global* minimizer is the point at which the function has the lowest value. These concepts are represented in Figure 4.21. A unique point may exist at which in one direction the point is a minimum and in another a maximum. Such a point is referred to as a *saddle* point (see Figure 4.20). Saddle points and local minimums can be avoided for large-signal parameter extraction if one of the optimization methods that follows is used. The concept of a minimum can be algebraically formalized as follows:

A point x^* is a strong minimizer of a function $F(\mathbf{x})$ if a scaler $\delta > 0$ exists such that $F(\mathbf{x}^*) < F(\mathbf{x}^* + \Delta\mathbf{x})$ for all $\Delta\mathbf{x}$ such that $0 < \|\Delta\mathbf{x}\| \leq \delta$.

A point is a weak minimizer of a function $F(\mathbf{x})$ if it is not a strong minimizer and a scaler $\delta > 0$ exists such that $F(\mathbf{x}^*) \leq F(\mathbf{x}^* + \Delta\mathbf{x})$ for all $\Delta\mathbf{x}$ such that $0 < \|\Delta\mathbf{x}\| \leq \delta$.

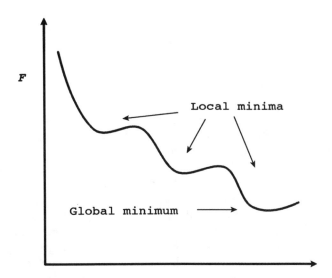

Figure 4.21 Local and global minimums.

The necessary conditions to define minimizers can be formulated. If $F(\mathbf{x})$ has continuous second derivatives, then it can be approximated at an arbitrary point $\mathbf{x} + \Delta\mathbf{x}$, given $F(\mathbf{x})$, $\mathbf{g}(\mathbf{x})$, and $\mathbf{G}(\mathbf{x})$, by using the first terms of the Taylor expansion:

$$F(\mathbf{x} + \Delta\mathbf{x}) = F(\mathbf{x}) + \Delta\mathbf{x}^T\mathbf{g}(\mathbf{x}) + \tfrac{1}{2}\Delta\mathbf{x}^T\mathbf{G}(\mathbf{x})\Delta\mathbf{x} + \cdots \qquad (4.69)$$

The higher order terms may be neglected if $\|\Delta\mathbf{x}\|$ is small. Using the first two terms of (4.69), the first-order necessary condition for a minimizer can be derived. If $\Delta\mathbf{x}^T\mathbf{g}(\mathbf{x}) < 0$, this implies that $F(\mathbf{x} + \Delta\mathbf{x}) < F(\mathbf{x})$, which contradicts the previous definition of a minimizer. The definition is still contradicted if $\Delta\mathbf{x}^T\mathbf{g}(\mathbf{x}) > 0$. Thus requiring $\Delta\mathbf{x}^T\mathbf{g}(\mathbf{x}) = 0$ is reasonable, which implies that

$$\mathbf{g}(\mathbf{x}^*) = 0 \qquad (4.70)$$

This is the first-order necessary condition for a minimizer. Equation (4.70) can also be satisfied by a maximum. Therefore, a second-order necessary condition is sought. If the first three terms of (4.69) are used at a strong minimum, the condition $\Delta\mathbf{x}^T\mathbf{G}(\mathbf{x}^*)\Delta\mathbf{x} > 0$ will ensure that $F(\mathbf{x} + \Delta\mathbf{x}) > F(\mathbf{x})$. This condition is true only if $\mathbf{G}(\mathbf{x}^*)$ is *positive definite*. A matrix is positive definite if its eigenvalues are all greater than zero. A matrix is *positive semidefinite* if its eigenvalues are greater than or equal to zero. A point can still be a strong minimum if $\Delta\mathbf{x}^T\mathbf{G}(\mathbf{x}^*)\Delta\mathbf{x} = 0$ and the third term in (4.69) is positive. The second-order necessary condition can then be expressed as

$$\Delta x^T G(x^*) \Delta x \geq 0 \qquad (4.71)$$

which implies that $G(x^*)$ must be positive semidefinite.

All of the gradient-based methods presented here are based on a quadratic model. These methods have been derived to work precisely if applied to a quadratic function. Yet these methods will have equal success if applied iteratively to generally well-behaved functions. If both the first and second derivatives are available, then the classical Newton method can be formulated, although often it cannot be used without modification due to its inability to deal with local minimums and saddle points. A similar model is used for restricted step methods such as that conceived by Levenberg and Marquardt. If second derivatives are not available, then they can be estimated in various ways. From this group emerges the quasi-Newton methods. When no derivatives are available, they can be estimated using finite differences with good success depending on the minimization method used. Using equation (4.69), we see that for a quadratic function the first three terms are exact; therefore, any function with a continuous second derivative approximates a quadratic function in a sufficiently small region near a minimum (maximum). Any method based on a quadratic model, therefore, will converge rapidly to a solution in the neighborhood of a minimum for a general nonlinear function. All Newton methods are based on this conjecture.

Convergence rates are important when considering the *robustness* of a particular method. Although this is a good indication of the worthiness of a particular method, it is no guarantee of good practical performance. The combination of experimentation and the rate of convergence test is the most reliable indication of good performance. If the error $x^k - x^*$ exists, then the *p'th order rate of convergence* is defined by the limit:

$$\beta = \lim_{k \to \infty} \frac{\|x^{k+1} - x^*\|}{\|x^k - x^*\|^p} \qquad (4.72)$$

where x^k represents the k'th iteration and x^* is the minimum. The term $\beta(0 \leq \beta < 1)$ is known as the *convergence ratio*. The most important cases are: $p = 1$ (*first order* or *linear convergence*) and $p = 2$ (*second order* or *quadratic convergence*).

Another important metric is the *condition number* given by

$$\gamma = \frac{\lambda_{max}}{\lambda_{min}} \qquad (4.73)$$

where λ_{max} and λ_{min} are the maximum and minimum eigenvalues of the Hessian matrix at the minimum. This is the ratio of the maximum to minimum curvature in two directions at the minimum. Usually, the larger γ is, the harder it becomes to find a minimum. When γ is small or the eigenvalues are grouped close to unity, then the problem is *well scaled*. Theoretically, when the condition number becomes an obstacle, a scaling technique called *preconditioning* can be implemented [21].

The most important part of setting up an optimization problem is defining the objective function. The choice of objective function affects both the numerical efficiency of the routine and, more importantly, the validity of the final solution. For large-signal parameter extraction, the primary concern is to match measured data to modeled data. Which measured data to use, however, is not obvious. The dc data, small-signal RF data, load-pull contours, large-signal S-parameters, and harmonic signal content as a function of input power have all been used by some investigators.

An example error function (objective function) is formulated and optimized later in this section. The formulated function uses measured and modeled drain current, transconductance, and output conductance. These quantities can be easily measured and are easily modeled using any large-signal model's drain current expression and derivatives. The function is then minimized using an appropriate optimization method. The error is defined in the form of a sum of the squares of nonlinear functions:

$$F(\mathbf{x}) = \sum_{i=1}^{m} \left[\sum_{j=1}^{p} r_{i,j}^2(\mathbf{x}) \right] \text{ for } i = 1(1)m, j = 1(1)p \qquad (4.74)$$

where $r_{i,j}$ represents the difference between measured and modeled drain current, transconductance, and output conductance; m is the total number of data points; p is the number of nonlinear device properties to be optimized (i.e., drain current, transconductance, output conductance, *et cetera*), and $\mathbf{x}$ is the group of n variables (model parameters such as α, β, V_{TO}, *et cetera*) to be optimized. The $r_{i,j}$ are known as *residuals* and are expressed as

$$r_{i,j} = y_{i,j} - M(\mathbf{x}, \mathbf{z}_i) \text{ for } i = 1(1)m, j = 1(1)p \qquad (4.75)$$

where $y_{i,j}$ represents the measured properties and $M(\mathbf{x}, \mathbf{z}_i)$ represents the model equations, which are a function of both the model parameters $\mathbf{x}$ and terminal voltages $\mathbf{z}_i$.

Varying importance can be placed on the nonlinear device properties by applying scalar quantities called *weighting factors*. The resulting function to be minimized is then given as

$$F(\mathbf{x}) = \sum_{i=1}^{m} \left[\sum_{j=1}^{p} \mathbf{w}_j r_{i,j}^2(\mathbf{x}) \right] \text{ for } i = 1(1)m, j = 1(1)p \qquad (4.76)$$

where the scalars w_j represent the weighting factors. The minimization of functions of this kind is called *weighted nonlinear least squares* [22]. The objective function can be formulated in many other ways; however, the form presented works well in practice.

Another point that should not be overlooked is the feasibility of determining the analytical partial derivatives of the objective function, which includes nonlinear functions such as device model equations. These derivatives are needed in equations (4.65) and (4.66). Determining analytical derivatives is an extremely tedious and time-consuming process that can be subject to errors. For this reason, finite differences can be used to approximate both the first and second partial derivatives. Because these algorithms will be incorporated into computer programs, machine tolerances such as *round-off errors, truncation errors,* and machine accuracy must be addressed [23, 24]. Nongradient-based optimization methods will also circumvent these problems.

4.3.3   Newton Methods

Newton methods can be used for optimizing an arbitrary function $F(x)$ of n independent variables, which is an iterative process. The k'th iteration consists of computing a *search vector* (descent direction) s^k from which the new iteration is obtained according to

$$x^{k+1} = x^k + \alpha^k s^k \tag{4.77}$$

where α^k is a scalar quantity varied according to some line search technique. A line search method is employed when proceeding from a point that is outside the region of convergence of the particular method of optimization in use. The line search seeks to find α^k to minimize $F(x^{k+1} + \alpha s^k)$ with respect to α before applying (4.77) [21]. Figure 4.22 illustrates the line search problem in one dimension.

The Taylor series of the gradient of a quadratic function at $x^{k+1} = x^k + s^k$ is

$$g^{k+1} = g(x^k + s^k) = g^k + Gs^k \tag{4.78}$$

with higher order terms equal to zero. If the minimum is at x^{k+1}, then g^{k+1} is zero and s^k is found by solving the system of equations given by

$$G^k s^k = -g^k \tag{4.79}$$

Newton's method will find the minimum of a quadratic function in exactly one iteration. If the function is not exactly quadratic, then the next iteration is given by (4.77) with $\alpha^k = 1$. The k'th iteration for Newton's method can be written

(1) Solve $G^k s^k = -g^k$ for s^k (solve system of equations).
(2) Set $x^{k+1} = x^k + \alpha^k s^k$ with α^k determined from a line search of choice.

Because (4.69) has a unique minimizer only when $G(x)$ is positive definite, Newton's method is well defined only for this condition. In general, nonlinear functions associated with large-signal parameter extraction do not have a positive definite Hessian matrix everywhere. This is one reason why Newton's method is not the best choice for solving large-signal parameter extraction problems. Two classes of methods that do guarantee a positive definite Hessian matrix, and thus global convergence, are the *quasi-Newton* or *variable metric* class of methods and the *Levenberg-Marquardt* or *restricted step* class of methods.

Quasi-Newton methods are very similar to the basic Newton method with the exception that the Hessian is approximated by a symmetric positive definite matrix. Equation (4.79) can be written as

$$s^k = -g^k G^{(k)-1} \tag{4.80}$$

The approximation matrix H^k replaces $G^{(k)-1}$ in (4.80) and is updated during each iteration. The classes of quasi-Newton methods differ in the formula used to update H^k. The k'th iteration of a quasi-Newton method can be written:

(1) Set $s^k = -g^k H^{(k)}$.
(2) Set $x^{k+1} = x^k + \alpha^k s^k$ with α^k determined from a line search of choice.
(3) Update H^k resulting in H^{k+1}.

Fletcher [25] suggests using the identity matrix I as the first estimate of H^k, although any symmetric positive definite matrix could be used. The appeal of quasi-Newton methods is that they require only first derivatives and s^k is obtained without solving a system of equations every k'th iteration. One of the most popular

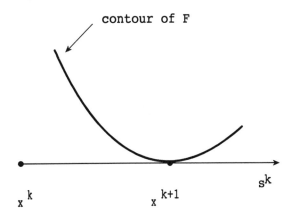

Figure 4.22 Line search in one dimension.

and widely used updating methods is the BFGS updating formula [22]. The BFGS formula is given by defining the differences:

$$\delta^k = \alpha^k s^k = x^{k+1} - x^k \tag{4.81}$$

$$\gamma^k = g^{k+1} - g^k \tag{4.82}$$

then

$$H^{k+1} = \left[I - \frac{\delta^k \gamma^{(k)T}}{\delta^{(k)T}\gamma^k} \right] H^k \left[I - \frac{\delta^k \gamma^{(k)T}}{\delta^{(k)T}\gamma^k} \right]^T + \frac{\delta^k \delta^{(k)T}}{\delta^{(k)T}\gamma^k} \tag{4.83}$$

Another class of methods is the restricted step methods, so called because they restrict the length of the step x^{k+1}. This class of methods is also derived from the basic Newton method. The search direction of the basic Newton method (4.80) is modified by giving it a bias toward the steepest descent vector $-g^k$ [25]. The new system of equations follows:

$$(G^k + \mu^k I)s^k = -g^k \tag{4.84}$$

where $\mu \geq 0$ and I is of order n. The function of the scalar μ is to control the length of the step and to force the matrix $(G^k + \mu I)$ to be positive definite. If $\mu = 0$, then the full Newton step is taken. Also note that decreasing μ corresponds to increasing $\|s^k\|$. The matrix $(G^k + \mu I)$ can be forced positive definite by setting $\mu \geq \|-\lambda_{min}\|$. We can see that μ can be changed in an iterative manner to obtain the desired results.

When a restricted step technique is applied to a nonlinear least-squares problem, it is known as the Levenberg-Marquardt method [26, 27]. Recalling that least-squares problems are defined in terms of the L_2 norm, derivatives of $F(x)$ are given by

$$g(x) = 2Jr \tag{4.85}$$

and

$$G(x) = 2JJ^T + 2\sum_{i=1}^{m} r_i \nabla^2 r_i \tag{4.86}$$

where $r_i = r_i(x)$ (the residuals) and J is the $n \times m$ *Jacobian matrix*, with n being the number of parameters and m the number of data points. The Hessian can be simplified to

$$G(x) = 2JJ^T \tag{4.87}$$

for small residual problems. Using (4.87), the restricted step method discussed earlier becomes

$$(\mathbf{JJ}^{(k)T} + \mu^k\mathbf{I})\mathbf{s}^k = -\mathbf{J}^k\mathbf{r}^k \text{ for } \mu \geq 0 \tag{4.88}$$

The basic algorithm for solving a nonlinear least-squares problem using (4.88) and (4.77) with $\alpha = 1$ is determined by how μ^k is chosen for each iteration. Marquardt's approach to the algorithm is as follows.

Prototype Levenberg-Marquardt Method Algorithm

    (1) start at some $\mathbf{x}^0$, set tolerance
    (2) set $\mu^0 = 0.01$, $v = 10$
    (3) for $k = 0,1,2, \ldots$,loop
    (4)   set $\mu^k = \mu^k/v$
    (5)   loop
    (6)   solve $(\mathbf{JJ}^{(k)T} + \mu^k\mathbf{I})\mathbf{s}^k = -\mathbf{J}^k\mathbf{r}^k$ for $\mathbf{s}^k$
    (7)   set $\mathbf{x}^{k+1} = \mathbf{x}^k + \mathbf{s}^k$
    (8)   if $F^{k+1} > F^k$, then
    (9)   set $\mu^k = \mu^k v$
   (10)   end if
   (11)   until $F^{k+1} < F^k$
   (12) set $\mu^{k+1} = \mu^k$
   (13) until $\|2\mathbf{J}^{(k+1)T}\mathbf{r}^{k+1}\| <$ tolerance.

Many variations are available for this basic algorithm, one of which will be demonstrated in the material on large-signal optimization using small-signal S-parameters.

    The criterion for terminating any of the above optimization methods is theoretically when $\mathbf{g}(\mathbf{x}) = 0$. However, because these algorithms are executed on digital computers with finite word lengths, this criterion can only be approximated to a certain precision. Another popular criterion is to check the difference in the function from one iteration to the next. If the difference is less than some tolerance for a number of consecutive iterations, then the program is terminated.

4.3.4 Random Optimization Techniques

One of the easiest optimization routines to implement is a simple random optimization method. Simple methods start by *bracketing* the minimum of the objective function, which requires a knowledge of the relative magnitudes of the individual variables to be minimized. This is usually possible for engineering applications such as parameter extraction of large-signal model parameter values. Elementary methods do not use gradient information of the objective function, and they may or may not converge to a minimum depending on the initial estimates of the brackets. The appeal of a random procedure lies in the ability to transform an algorithm into a computer program in only a few hours. They can give some qualitative information about the problem to be optimized; such information is often all that is needed for a particular application. An example algorithm is given below.

Random Optimization Method Algorithm

(1) set N = number of variables
(2) set tol = routine termination value
(3) bracket each variable by determining a maximum and minimum value for each variable
(4) loop
(5)   set n = random number between 0 and 1
(6)   for i = 1 to N
(7)   set $x_i = x_{min} + (x_{max} - x_{min})*n$
(8)   next i
(9)   evaluate $F^k(\mathbf{x}) \rightarrow F^{k+1}(\mathbf{x})$
(10)   if $F^{k+1}(\mathbf{x}) < F^k(\mathbf{x})$, then save $\mathbf{x}$
(11) until $F^{k+1}(\mathbf{x})$ < tol.

This method is often used as a preamble to a gradient-based method which allows the gradient method to start at a new point every time it is run. This addresses the problem of local minima, which sometimes cause problems in gradient-based methods. A number of techniques can be employed in order to make the routine more sophisticated. A self-bracketing routine is helpful. Such a routine updates the original bracket each time a better $\mathbf{x}$ is found. A simple bracket updating procedure is to choose the new brackets around a percentage of the new iterate x_i. Note that simple random methods only minimize a function without the guarantee of finding a global minimum. A more sophisticated random-based method that avoids entrapment in local minima is *simulated annealing*.

Simulated annealing has been used with excellent results to solve extremely large optimization problems such as the traveling salesman problem [28] and for modeling microwave semiconductor devices [29]. The method of simulated annealing is based on a random process and closely resembles the annealing techniques utilized in the manufacture of semiconductors. At high temperatures, atoms move freely in molten material. When the material is cooled quickly, the atoms lose their thermal mobility and never reach a well-ordered low energy crystalline state. Annealing is the process of slow cooling of a material, which allows the atoms to distribute themselves in a minimum energy state. The method of simulated annealing has features that mimic this natural process using a Boltzmann-like law to determine the probability of accepting a solution for an iteration k. For data fitting purposes, a least-squares objective function such as that of (4.76) is appropriate. The predominant improvement of simulated annealing over the conventional heuristic methods is the criterion for accepting a new parameter vector $\mathbf{x}$. The new parameter vector $\mathbf{x}$ is accepted, as in the previous methods, if the objective function is reduced. However, for cases in which the objective function increases, the new parameter vector still has a finite probability of being accepted.

The probability function P can be defined

$$P(\mathbf{x}^k, T^k) = e^{[-F(\mathbf{x}^k)/T^k]} \tag{4.89}$$

where $F(\mathbf{x}) = F(\mathbf{x}^{k+1}) - F(\mathbf{x}^k)$, $\mathbf{x}^{k+1}$, and $\mathbf{x}^k$ are at the $k+1$ and k iteration step, and T^k is the pseudotemperature, which is artificially decreased during the simulated annealing process. Similar to an actual annealing process, an initial temperature and temperature schedule are used. The temperature schedule can be defined as $T^{k+1} = \alpha T^k$. Values of $T^0 \geq 500$ (initial temperature) and $\alpha = 0.90$ have been found to work adequately for some applications. The parameter vector $\mathbf{x}$ is enclosed using a bracketing scheme. A variance is applied to the parameter matrix $\mathbf{x}$ by using a random number R where $0 \leq R \leq 1$. If the new parameter value is greater than one of the bracket points, the parameter value is set to the bracket point.

Prototype Simulated Annealing Algorithm

(1) $T^0 = 700$
(2) $k = 0, j = 0$
(3) bracket $\mathbf{x}$ $\pm 20\%$ to $\pm 60\%$
(4) set $\mathbf{x}$ to starting values
(5) $\alpha = 0.9$
(6) tol $= 0.001$
(7) evaluate $F(\mathbf{x}^k)$
(8) loop
(9) $T^{j+1} = \alpha T^j$
(10) generate $(0 \leq R \leq 1)$
(11) $x_i^{k+1} = x_i^k + x_i^k(R/10)$
(12) evaluate $F(\mathbf{x}^k) \rightarrow F(\mathbf{x}^{k+1})$
(13) $F(\mathbf{x}^k) = F(\mathbf{x}^{k+1}) - F(\mathbf{x}^k)$
(14) if $F(\mathbf{x}^k) < 0$, then keep $\mathbf{x}^{k+1}$
(15) else

(16) $P(\mathbf{x}^{k+1}, T^{j+1}) = e[-F(\mathbf{x}^k)/T^{k+1}]$
(17)   generate $(0 \le R \le 1)$
(18)   if $P(\mathbf{x}^{k+1}, T^{j+1}) \ge R$, then keep $\mathbf{x}^{k+1}$
(19) $j = j + 1$
(20) end if
(21) $k = k + 1$
(22) until $F(\mathbf{x}^k) \le$ tol (five consecutive iterations).

The key benefits of using such an algorithm are its simplicity, and its ability to exit from a local minimum. Simulated annealing can sometimes be considered a viable alternative to numerically intensive algorithms based on Newton's method.

4.3.5   A Large-Signal Parameter Extraction Methodology

Various large-signal parameter extraction methods exist for determining the large-signal model parameters for MESFETs and HEMTs. Parameter extraction using large-signal measurements such as power measurements [30] and spectrum measurements [20] has been implemented. A more efficient method, however, is to use dc and small-signal S-parameter measurements to determine large-signal device behavior. The method of predicting nonlinear behavior of GaAs MESFET devices by determining the bias and frequency dependence of S-parameters was originally proposed by Willing *et al.* [31].

Figure 4.23 is the small-signal model of a MESFET or HEMT with the boxed circuit elements being the dominant nonlinear elements. These nonlinear element values can be determined by the direct extraction method presented in Section 4.2.

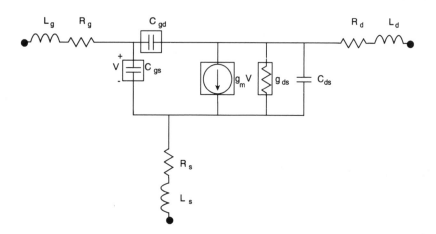

Figure 4.23 Small-signal model of a MESFET or HEMT with the dominant nonlinear elements boxed.

Large-signal device behavior is approximated by measuring S-parameters over many bias voltages from the linear through the saturation regions of device operation. Static dc measurements are also made at the bias points where the S-parameters are measured. The nonlinear equivalent circuit element values are extracted at all bias voltages, thereby creating an equivalent circuit for the device at each bias point. The choice of bias levels to consider using such a technique is determined by the particular device and application. These issues are discussed in more detail in Chapter 3. The parasitic resistances, inductances, and drain-source capacitance C_{ds} are considered linear elements and remain constant for all bias voltages. A compression of data is realized from the S-parameters to the equivalent circuit. The importance of the data compression is apparent when the parameter extraction problem is solved using explicitly defined optimization goals.

The technique described above offers several advantages over many other large-signal parameter extraction methods. It eliminates the need for time-consuming, expensive, and demanding measurements, while providing excellent agreement between measured and modeled characteristics. Load-pull measurements, pulsed I-V measurements, and transient or harmonic output power measurements all require significant expenditures of time or capital. The complexity of these measurements also often reduces their accuracy. The dc and S-parameter data required of this method are easily taken using equipment commonly available in any microwave laboratory. Automation of the measurements also requires minimal effort. A second advantage of the technique is that large amounts of data are compressed. This further simplifies the extraction process. Use of bias-dependent equivalent circuit values instead of direct S-parameter or harmonic power data can reduce the computational requirements for the parameter extractor by more than 1000 times. Such a reduction in parameter extraction complexity often leads to improved accuracy because more data can be considered. Finally, an extremely useful by-product of the procedure described above is that small-signal equivalent circuit models are obtained for a number of bias levels. These models can be scaled to describe devices with arbitrary gate widths and used in small-signal simulations.

At the heart of this method, based on characterization using dc and S-parameter measurements, is an optimization routine. Success can be achieved with a routine that utilizes both a random method and a variation of the modified Newton method of Levenberg and Marquardt. The extraction problem is defined by first obtaining the dc measurements and nonlinear equivalent circuit elements over bias from measured S-parameters. An efficient method for accomplishing this task is presented in Section 4.2. The device model parameters are then adjusted by using the optimization program. This fits the measured nonlinear elements g_m and g_{ds} ($1/r_{ds}$), and the measured dc drain-source current I_{ds} to the modeled values obtained using a device model. Using this technique, the device model can be used to simulate both the dc and frequency-dependent large-signal characteristics. In a similar manner, the nonlinear gate-source C_{gs} and gate-drain C_{gd} capacitance parameters can be determined using a capacitance model of choice. The extraction of the capacitance-voltage model parameters is therefore independent of the extraction of other model parameters. This further simplifies the problem.

The model equations are incorporated into the extraction programs for each model. The drain current, transconductance, and output conductance should be represented with analytical expressions within the optimizer. These equations are presented in Section 2.3. For the models presented in Chapter 2, the basic model equation for drain current is used to predict both dc and ac responses. A uniquely defined objective function is used to measure simultaneously the error in the dc and ac characteristics of the device. The objective function is termed the error magnitude unit or EMU and is defined by equation (4.90). This formula is typical of a nonlinear weighted least-squares problem.

$$\text{EMU} = \frac{\text{EMU}'}{3n} \tag{4.90}$$

where n = number of bias points and

$$\text{EMU}' = \sum_{i=1}^{n} [E_{Ids}(i)WF_{Ids} + E_{gm}(i)WF_{gm} + E_{gds}(i)WF_{gds}]100$$

and

$$E_{\alpha\alpha\alpha}(i) = \frac{[\text{meas}(i) - \text{mod}(i)]^2}{\text{meas}(i)^2}$$

with WF_{Ids}, WF_{gm}, and WF_{gds} as normalized weighting factors for I_{ds}, g_m, and g_{ds}, respectively. The ac characteristics are described by fitting the modeled transconductance and output conductance to measured values of microwave transconductance and output conductance. The dc characteristics are described by fitting the modeled drain-source current behavior.

Note that none of the large-signal models discussed in Chapter 2 accounts for the low frequency dispersion in output conductance and transconductance of the device. Therefore, the error function of equation (4.90) cannot achieve a value of zero because both dc and RF error are considered in this formula. Other parameter extraction routines based on other measurements can be utilized. Regardless of the extraction technique utilized, however, models that do not account for low frequency dispersion cannot simultaneously match both dc and RF characteristics of the device. By using both dc and RF characteristics along with the described weighting functions, an acceptable compromise can be obtained from most of the large-signal models presented.

The optimization routine is comprised of a two-stage method. Device model parameters are bracketed first and optimized using a simple random technique such as that described earlier, followed by the Levenberg-Marquardt (L-M) method. The use of a random technique first enables the L-M optimization to begin from a different starting point each time the extraction program is executed, thus addressing the problem of entrapment in local minima.

The L-M method uses both first and second derivatives. In terms of the device models and the EMU, these derivatives are expressed as

$$
\mathbf{g}(\mathbf{p}) = \begin{bmatrix} \partial \text{EMU}'/\partial p_1 \\ \partial \text{EMU}'/\partial p_2 \\ \vdots \\ \partial \text{EMU}'/\partial p_n \end{bmatrix} = \nabla \text{EMU}'(\mathbf{p}) \tag{4.91}
$$

$$
\mathbf{G}(\mathbf{p}) = \begin{bmatrix} \partial^2 \text{EMU}'/\partial p_1^2 \ldots \partial^2 \text{EMU}'/\partial p_1 \; \partial p_n \\ \cdot \qquad\qquad \cdot \quad \cdot \\ \cdot \\ \partial^2 \text{EMU}'/\partial p_n \partial p_1 \ldots \quad \partial^2 \text{EMU}'/\partial p_n^2 \end{bmatrix} \tag{4.92}
$$

where $\mathbf{p}$ is the model parameter vector, which contains the values for the n model parameters. Because the derivatives of the EMU are extremely difficult to formulate analytically, they are obtained using finite differences. A prototype algorithm for a large-signal parameter extraction computer program based on the L-M method is presented below:

Prototype Parameter Extraction Algorithm

(1) loop
(2)   calculate $\mathbf{g}(\mathbf{p})$, $\mathbf{G}(\mathbf{p})$
(3)   test eigenvalues λ_i of $\mathbf{G}(\mathbf{p})$ (Jacobi method)
(4)   loop
(5)   start flag = 1
(6)   solve $[\mathbf{G}^k(\mathbf{p}) + \mu^k \mathbf{I}]\mathbf{s}^k = -\mathbf{g}(\mathbf{p})$ (Gauss-Jordan)
(7)   temporarily set $\mathbf{p}^{k+1} = \mathbf{p}^k + \mathbf{s}^k$
(8)   calculate $\text{EMU}(\mathbf{p}^{k+1})$
(9)   if $\text{EMU}(\mathbf{p}^{k+1}) > \text{EMU}(\mathbf{p}^k)$ or any $\lambda_i < 0$, then
(10)   adjust L-M parameter μ (see prototype algorithm)
(11)   else
(12)   set $\mathbf{p}^{k+1} = \mathbf{p}^k + \mathbf{s}^k$
(13)   end if

(14)   until $\text{EMU}(\mathbf{p}^{k+1}) < \text{EMU}(\mathbf{p}^k)$
(15) until term flag or $\text{EMU}(\mathbf{p}^{k+1}) - \text{EMU}(\mathbf{p}^k) < 0.001$ (three consecutive iterations).

Prototype L-M Parameter Adjustment Algorithm

(1) $v = 10$
(2) if start flag then
(3) $\text{LM}v = le30$ (some large value)
(4)   if $\|\lambda_i\| < \text{LM}v$ and $\lambda_i \neq 0$, then $\text{LM}v = \lambda_i$ (where λ_i is the smallest negative eigenvalue)
(5) $\text{LM}v = \|\text{LM}v\|$
(6)   start flag $= 0$
(7) else
(8) $\text{LM}v = v * \text{LM}v$
(9) end if
(10) set $\mathbf{G}_{i,j}(\mathbf{p}) = \text{LM}v + \mathbf{G}_{i,j}(\mathbf{p})$ where $i = j$
(11) return.

The robustness of the parameter extraction algorithm can be tested by running the extractor program multiple times from different starting points and examining the effect this has on the final solution. The results of such a test for the advanced Curtice model (Chapter 2) are illustrated in Figure 4.24. The individual parameters of the model are perturbed, one at a time, by $\pm 20\%$ to $\pm 50\%$ of an optimal value. After perturbation, the parameter extractor is restarted. This determines whether or not the parameters are converging to a unique solution. The L-M optimization process is allowed to terminate itself, thus determining the optimum parameter values. The criterion for termination is established to be when the EMU does not decrease by 0.001 for three consecutive iterations. During this test, the other parameters are started from the value reached after the random optimization operation. Similar results to those illustrated in Figure 4.24 are obtained for the other parameters.

Another test can be performed in which EMU *versus* each parameter is graphed and examined to determine if the objective function is well behaved in each parameter space. For most of the large-signal models presented in Chapter 2, the objective function is found to be smooth and continuous with no observable local minima. The algorithm described above has been used extensively for parameter extraction of MESFET and HEMT models and is found to be very robust and efficient.

The EMU of equation (4.90) is used to determine all model parameter values except capacitance parameters. As stated previously, capacitance parameters can

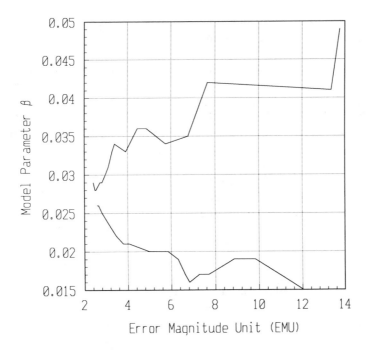

Figure 4.24 Computed value for one model parameter as a function of the resulting EMU determined for several optimization runs.

be determined independently for most models using the same concepts presented earlier in the section. A capacitance EMU is easily defined by

$$EMU_{cap} = \frac{EMU'_{cap}}{2n} \qquad (4.93)$$

where n = number of bias points and

$$EMU'_{cap} = \sum_{i=1}^{n} [E_{Cgs}(i)WF_{Cgs} + E_{Cgd}(i)WF_{Cgd}]100$$

and

$$E_{C\alpha\alpha}(i) = \frac{[meas(i) - mod(i)]^2}{meas(i)^2}$$

with WF_{Cgs} and WF_{Cgd} as normalized weighting factors. Optimization of the capacitance parameters is then accomplished by minimizing (4.93) by means of any of the optimization schemes described.

Figures 4.25 through 4.30 illustrate the measured *versus* modeled results of dc current-voltage characteristics, RF output conductance, and RF transconductance using this optimization approach with the weighting factors all set to 1.0 for both a MESFET and a HEMT. The devices used in these measurements are a depletion-mode RF 0.5×300-μm MESFET that can be probed and a 0.7×200-μm pseudomorphic HEMT. The parameter extraction is performed for the advanced Curtice model for the MESFET and the advanced Curtice model with modifications for the HEMT (see Chapter 2). Note that better agreement between the measured and modeled characteristics shown in Figures 4.25, 4.26, or 4.27 can be obtained at the expense of the other two. Exact agreement of all three simultaneously, however, is not possible with any of the models considered. This limitation of the models is caused by their inability to model the low frequency dispersion of output conductance and transconductance observed in GaAs MESFETs as described in Section 1.4.

Several comments need to be made regarding the parameter extraction process for physically based models. Because physically based models derive electrical performance predictions from physical dimensions and material properties of the

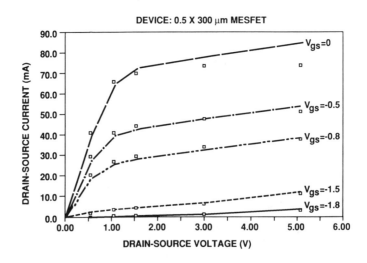

Figure 4.25 A comparison of dc measured and modeled drain-source current *versus* drain-source voltage. Blocks are measured for a 0.5×300-μm MESFET, while the solid lines are predictions from the advanced Curtice model using optimized parameters.

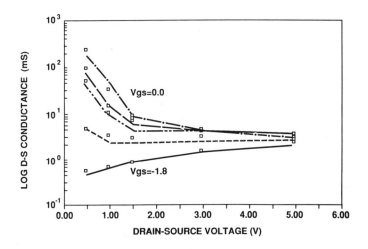

Figure 4.26 A comparison of RF measured and modeled output conductance. Blocks are measured for a 0.5×300-μm MESFET, while the solid lines are predictions from the advanced Curtice model using optimized parameters.

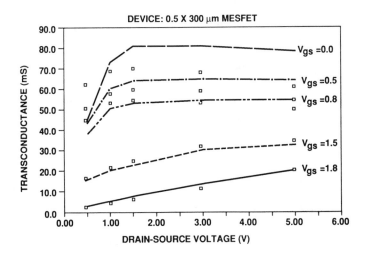

Figure 4.27 A comparison of RF measured and modeled transconductance. Blocks are measured for a 0.5×300-μm MESFET, while the solid lines are predictions from the advanced Curtice model using optimized parameters.

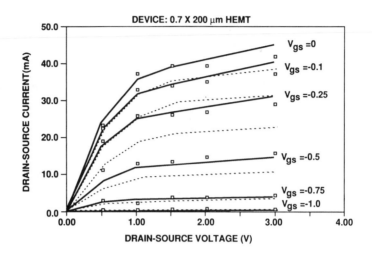

Figure 4.28 A comparison of dc measured and modeled drain-source current versus drain-source voltage. Blocks are measured for a 0.7×200-μm HEMT. Solid lines are predictions from the advanced Curtice model with HEMT modifications using optimized parameters. Dashed lines are predictions from the advanced Curtice MESFET model using optimized parameters.

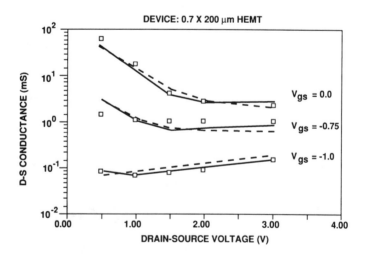

Figure 4.29 A comparison of RF measured and modeled output conductance. Blocks are measured for a 0.7×200-μm HEMT. Solid lines are predictions from the advanced Curtice model with HEMT modifications using optimized parameters. Dashed lines are predictions from the advanced Curtice MESFET model using optimized parameters.

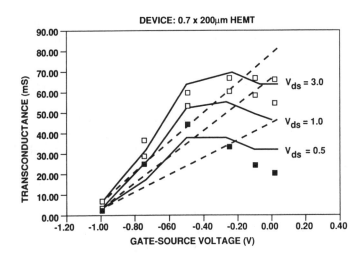

Figure 4.30 A comparison of RF measured and modeled transconductance. Blocks are measured for a 0.7 × 200-μm HEMT. Solid lines are predictions from the advanced Curtice model with HEMT modifications using optimized parameters. Dashed lines are predictions from the advanced Curtice MESFET model using optimized parameters.

device, parameter extraction amounts to measuring these physical characteristics. In theory, no electrical measurements (dc, RF, or large signal) are required. This is one of the major advantages potentially offered by physical models. Such a view, however, overlooks several practical issues.

Parameters typically required of a physically based MESFET or HEMT model include semiconductor layer thicknesses, gate dimensions, dimensions of other surface features such as terminal spacings, doping densities, and transport properties such as electron mobility and saturated velocity. Unfortunately for the circuit designer, device manufacturers and foundries typically consider such information proprietary. Thus, the physical characteristics of the device are not often known when the design process begins. To measure surface geometry features directly, a scanning electron microscope (SEM) image can be exploited. Measurement of layer thickness and doping densities requires more specialized equipment. Combinations of capacitance-voltage measurements, deep level transient spectroscopy (DLTS) measurements, or Hall effect measurements [32] can be employed. These measurements often require the use of special test structures. In many large laboratories (especially semiconductor processing facilities), access to such equipment is possible, while in smaller facilities such measurements are difficult to obtain. Even when such measurements can be made, the accuracy obtainable is often not sufficient for modeling requirements. For example, perturbations of only several angstroms in the value of the spacer layer thickness used in a HEMT model will have a significant effect on performance predictions. Yet measurements of this thickness are extremely difficult to make with an accuracy that even begins to approach this requirement. Likewise, gate length and other layer dimensions for both the MESFET and the HEMT need to be determined to extreme accuracy if acceptable performance predictions are to be obtained from a physically based model.

An attractive alternative to making direct measurements of the physical characteristics of a finished device is to make electrical measurements of device behavior and then extract physical properties. This is easily accomplished using the parameter extraction algorithm described previously. Using this technique, the physical dimensions and material properties of the device are treated as model parameters. The parameter extractor effectively measures the physical characteristics of the device using standard electrical measurement techniques. The resulting model still has the benefits of a physical model (it can be scaled easily or used for yield predictions, for example) but can also be implemented easily. This technique can be used as a powerful tool for device design or as a process monitor because all device processes can be examined with a single set of easy-to-make electrical measurements.

The weighting factors contained in the optimization objective function can be used to apply a bias toward either the dc or RF characteristics during the parameter extraction process. A model whose parameters have been chosen by this process will show a prejudice in the device performance predictions depending on the weighting scheme chosen. A bias toward the dc characteristics could be used, for example, when using a FET as a current source in a MMIC circuit. The design of such a MMIC circuit requires accurate dc as well as RF predictions. The dc predictions, however, may be more consequential in this case. To account for this importance, the optimization process can be more heavily weighted toward dc I_{ds} than RF g_m or g_{ds}. Figures 4.31 through 4.34 illustrate these concepts. The figures show the agreement obtained between measured and modeled device characteristics for three different weighting schemes. The parameter extraction scheme described above was applied using the advanced Curtice model of Chapter 2 but with the following weighting schemes:

(1) dc weighting set $WF_{Ids} = 1.0$, $WF_{gm} = WF_{gds} = 0.0$
(2) RF weighting set $WF_{Ids} = 0.0$, $WF_{gm} = WF_{gds} = 1.0$
(3) dc/RF weighting set $WF_{Ids} = WF_{gm} = WF_{gds} = 1.0$

The results of optimizing to the dc and RF data are demonstrated in Figure 4.31, which shows measured and modeled output resistance r_{ds} as a function of V_{ds}. For the first weighting scheme, only the dc current is considered. As is evident from the graph, the model simulates the dc behavior fairly well. As a result of the frequency dependency of the resistance, however, the model predictions do not match the measured RF values well. In contrast, when the model is optimized to only the RF values of g_{ds} and g_m, the RF agreement is much better. A compromise between dc and RF behavior is obtained when the model is optimized to both the dc and RF behavior. Because two RF parameters are used, and only one dc parameter, the optimization is effectively weighted in favor of RF performance, which is evident in the results. A comparison of the modeled transconductance for each of the three optimization conditions reveals similar results, as illustrated in Figure 4.32.

The effect of considering only dc data on the model's RF accuracy is illustrated in Figure 4.33, in which the gain of the MESFET is plotted as a function of the available power at a frequency of 2.1 GHz. The dc-optimized model predicts significantly higher gain than measured. This is mainly a result of the low frequency dispersion of the device transconductance.

Very accurate microwave models may be extracted by optimizing the model performance using only the RF data. One purpose of large-signal parameter extraction, though, is to derive a model that is valid for all conditions, and RF optimization results in significant error in the dc current predictions, as demonstrated in

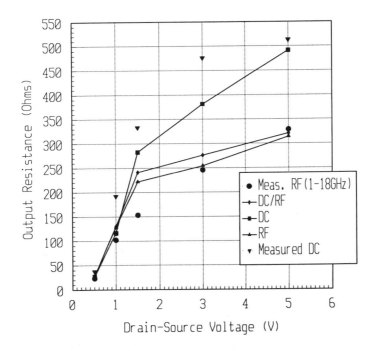

Figure 4.31 Measured and modeled microwave output resistance *versus* drain-source voltage using different weighting schemes for a 0.5 × 300-μm MESFET. The dc data correspond to WF_{Ids} = 1.0, WF_{gm} = WF_{gds} = 0.0; RF data correspond to WF_{Ids} = 0.0, WF_{gm} = WF_{gds} = 1.0; dc/RF data correspond to WF_{Ids} = WF_{gm} = WF_{gds} = 1.0. The WF factors are defined in equation (4.90) and the measurement frequency range is 1 to 18 GHz.

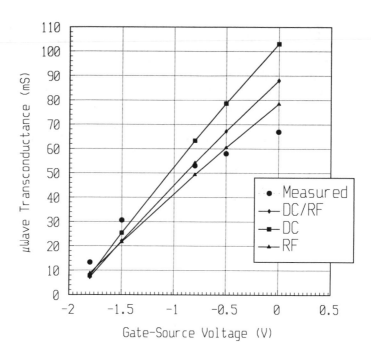

Figure 4.32 Measured and modeled microwave transconductance *versus* drain-source voltage using different weighting schemes for a 0.5 × 300-μm MESFET. The dc data correspond to WF_{Ids} = 1.0, WF_{gm} = WF_{gds} = 0.0; RF data correspond to WF_{Ids} = 0.0, WF_{gm} = WF_{gds} = 1.0; dc/RF data correspond to WF_{Ids} = WF_{gm} = WF_{gds} = 1.0. The WF factors are defined in equation (4.90) and the measurement frequency range is 1 to 18 GHz.

Figure 4.34. Optimization of the models to both the dc and RF performance provides a reasonable compromise between the two performance extremes for the models of Chapter 2.

In addition to predicting large-signal device performance, large-signal models can also be used to estimate small-signal S-parameters. This may be useful when measured S-parameters are only available for a limited number of bias points or in order to achieve data compression when storing data for a large number of devices. Small-signal S-parameters for the equivalent circuit of Figure 4.23 for both a MESFET and HEMT, at an amplifier bias of $\frac{1}{2}I_{dss}$, are shown in Figures 4.35 through 4.40. The figures show a comparison between measured S-parameters and those predicted from the advanced Curtice model (or the advanced Curtice model with HEMT degradation terms added), which has its parameters determined using the methods described above. All linear elements were obtained using the direct extraction method. The agreement between measured and modeled small-signal S-

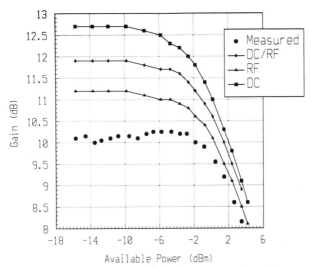

Figure 4.33 Measured and modeled gain *versus* available input power using different weighting schemes for a 0.5 × 300-μm MESFET. The dc data correspond to $WF_{Ids} = 1.0$, $WF_{gm} = WF_{gds} = 0.0$; RF data correspond to $WF_{Ids} = 0.0$, $WF_{gm} = WF_{gds} = 1.0$; dc/RF data correspond to $WF_{Ids} = WF_{gm} = WF_{gds} = 1.0$. The WF factors are defined in equation (4.90) and the measurement frequency is approximately 2.1 GHz.

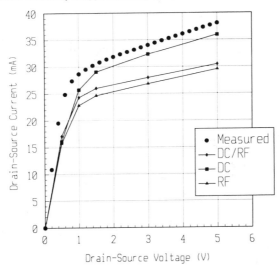

Figure 4.34 Drain-source current *versus* drain-source voltage using different weighting schemes for a 0.5×300-μm MESFET. The dc corresponds to $WF_{Ids} = 1.0$, $WF_{gm} = WF_{gds} = 0.0$; RF corresponds to $WF_{Ids} = 0.0$, $WF_{gm} = WF_{gds} = 1.0$; dc/RF corresponds to $WF_{Ids} = WF_{gm} = WF_{gds} = 1.0$.

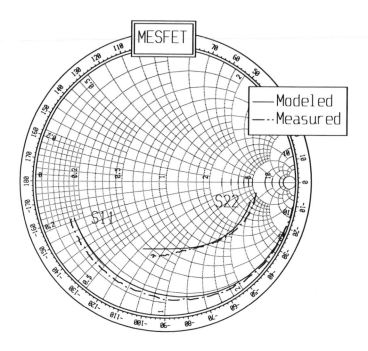

Figure 4.35 A comparison of measured *versus* modeled forward and reverse reflection coefficients S_{11} and S_{22}. The measured data are from a 0.5×300-μm MESFET biased at $1/2I_{dss}$. The modeled data are generated using the optimized large-signal model parameters of the advanced Curtice model. (Frequency range is 1–18 GHz.)

parameters is remarkable considering the number of data transformations that takes place through the entire parameter extraction process.

The accuracy of a large-signal model prediction depends in large part on how precisely the device model parameters can be extracted. The ultimate test of this accuracy is the ability of the resulting model to predict large-signal performance. One large-signal test of model accuracy can be made by comparing the measured and modeled power spectrum of a device as a function of input power level. This can be accomplished by using a single input signal or by using two signals that are at slightly different frequencies. Figure 4.41 illustrates two-tone measurement data and the results of simulations for a MESFET and three of the models presented in Chapter 2. The measurement is described in Chapter 3.

Using the parameter extraction technique described above, harmonic measurements were simulated for the MESFET using the advanced Curtice model, Statz model, and basic Curtice model (Chapter 2). The tests involved medium

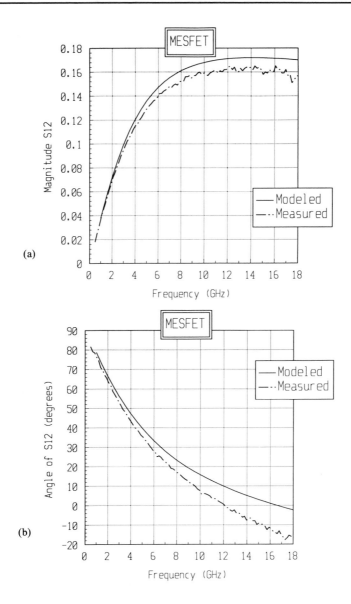

Figure 4.36 A comparison of measured and modeled S-parameter data. The measured data are from a 0.5×300-μm MESFET biased at $1/2I_{dss}$. The modeled data are generated using the optimized large-signal model parameters of the advanced Curtice model: (a) a comparison of the measured *versus* modeled magnitude of the reverse transmission coefficient S_{12}; (b) a comparison of the measured *versus* modeled angle of the reverse transmission coefficient S_{12}.

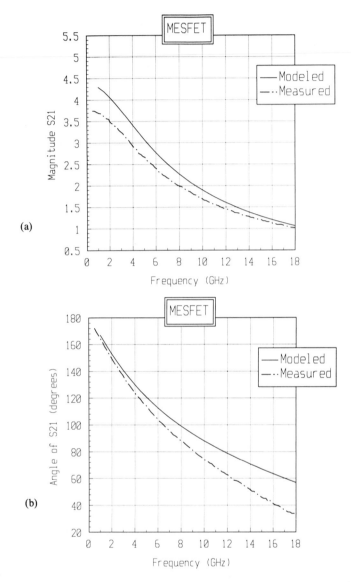

Figure 4.37 A comparison of measured and modeled *s*-parameter data. The measured data are from a 0.5×300-μm MESFET biased at $1/2I_{dss}$. The modeled data are generated using the optimized large-signal model parameters of the advanced Curtice model: (a) a comparison of the measured *versus* modeled magnitude of the forward transmission coefficient S_{21}; (b) a comparison of the measured *versus* modeled angle of the forward transmission coefficient S_{21}.

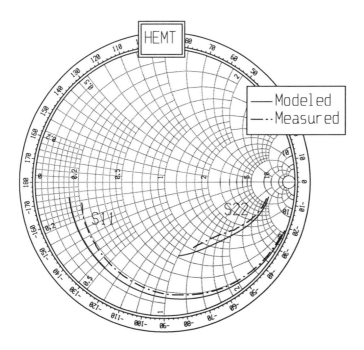

Figure 4.38 A comparison of measured *versus* modeled forward and reverse reflection coefficients S_{11} and S_{22}. The measured data are from a 0.7×200-μm HEMT biased at $1/2I_{dss}$. The modeled data are generated using the optimized large-signal model parameters of the advanced Curtice model with HEMT correction terms. (Frequency range is 1–18 GHz.)

power ion-implanted MESFETs. The recommended maximum drain-source bias level for these devices is 5 V. For this reason, device characterization was limited to this bias range. For other applications, this bias range may not be sufficient. The simulations were performed using a commercially available time-domain circuit simulator and results were also confirmed using a commercially available harmonic balance program. The two-tone response of the MESFET in a 50-Ω system was measured using equal-power signals at 2.10 and 2.12 GHz. The harmonic content of the output signal of the device was measured for a range of input power levels. All three models show agreement for the fundamental and third harmonic frequency. The advanced Curtice model shows the best agreement followed by the Statz model and basic Curtice model, respectively. The accuracy of these predictions corresponds exactly to the relative ranking of these models by the EMU presented in Section 2.3.

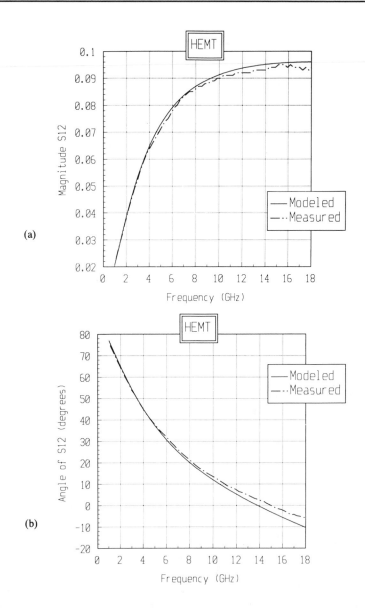

Figure 4.39 A comparison of measured and modeled *s*-parameter data. The measured data are from a 0.7×200-μm MESFET biased at 1/2I_{dss}. The modeled data are generated using the optimized large-signal model parameters of the advanced Curtice model with HEMT correction terms: (a) a comparison of the measured *versus* modeled magnitude of the reverse transmission coefficient S_{12}; (b) a comparison of the measured *versus* modeled angle of the reverse transmission coefficient S_{12}.

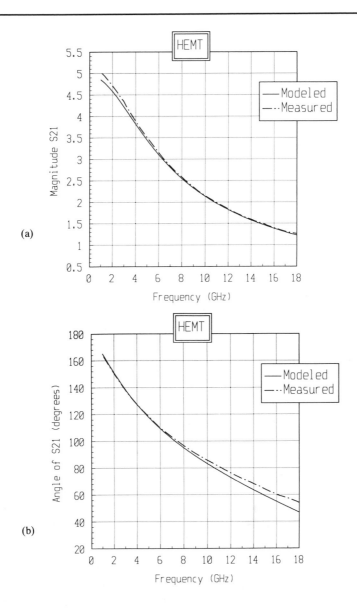

Figure 4.40 A comparison of measured and modeled *s*-parameter data. The measured data are from a 0.7×200-μm MESFET biased at $1/2I_{dss}$. The modeled data are generated using the optimized large-signal model parameters of the advanced Curtice model with HEMT correction terms: (a) a comparison of the measured *versus* modeled magnitude of the forward transmission coefficient S_{21}; (b) a comparison of the measured *versus* modeled angle of the forward transmission coefficient S_{21}.

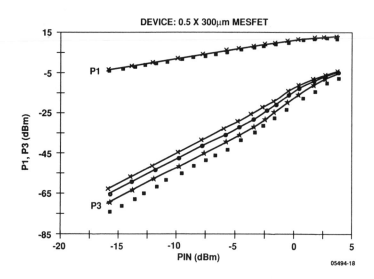

Figure 4.41 Fundamental and third harmonic power *versus* input power. The blocks are measured data: stars are predicted values from the advanced Curtice model; dots are predicted values from the Statz model; and x's are predicted values from the Curtice model.

Although the EMU for the Statz model is almost identical to that of the basic Curtice model, the Statz model shows better agreement to the measured data for both the fundamental and third harmonic frequencies. The inclusion of bias-dependent capacitances C_{gs} and C_{gd} significantly improves the prediction of the power of both the second and third harmonic frequencies [33]. Therefore, increasing the accuracy of the capacitance models should increase the complete large-signal model's ability to predict the amplitude of generated harmonics. The bias-dependent capacitance expressions for the Statz model are better than those associated with the basic Curtice model. This is the primary reason the Statz model shows a closer fit to the measured data. Figure 4.42 illustrates the third-order intercept point as a function of input power. The measured intercept point is +29 dBm.

The parameter extraction technique described above works well for many applications and can serve as the basis for more complex procedures. Other measurement data can also be incorporated into the EMU expression. This is necessary, for example, when the application of interest dictates that device breakdown is important. To determine appropriate breakdown voltage parameter values [see equation (2.125b)], some measurement of device breakdown characteristics is clearly required. As discussed in Chapter 3, this may be accomplished using pulsed measurement techniques. Likewise, the EMU expression must include the break-

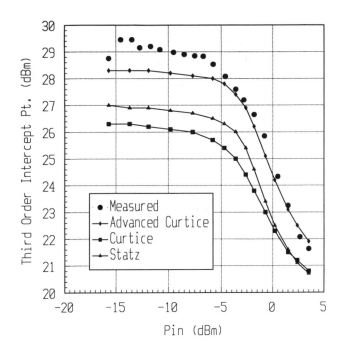

Figure 4.42 Measured and modeled third-order intercept point as a function of input power.

down terms. Other data, such as spectral output power as a function of input power, or power saturation can also be considered in the EMU.

4.4 NOISE MODEL EXTRACTION

Noise model extraction is the process of obtaining model parameters from measured noise data. Additional information describing the device, such as a small-signal model, is also needed for these models. The complexity of this process depends primarily on the method of characterization (Chapter 3), device noise model (Chapter 2), and, to some extent, the desired accuracy of the extracted parameters.

4.4.1 Direct Minimum Noise Figure Measurements

The direct minimum noise figure measurement technique described in Chapter 3 characterizes the noise properties of a MESFET or HEMT by applying a known admittance at the input of the device. The noise figure is monitored while adjusting the input impedance to achieve minimum noise figure. At this condition, the device exhibits minimum noise figure F_{min}, and the optimal source admittance is that which is applied at the device input (i.e., $Y_{opt} = Y_g$, and $F_{min} = F(Y_{opt})$. The equivalent noise resistance R_n is determined from

$$F(Y_g) = F_{min} + \frac{R_n}{G_g} [(G_g - G_{opt})^2 + (B_g - B_{opt})^2] \tag{4.94}$$

where

F_{min} = minimum value of F with respect to Y_g,
$Y_{opt} = G_{opt} + jB_{opt}$ = admittance value at which $F = F_{min}$,
R_n = equivalent noise resistance, and
Y_g = admittance connected at the device input.

For this characterization method, noise parameter extraction is trivial, assuming proper calibration of the equipment. However, as discussed in Chapter 2, this method can be inaccurate due to the requirement for a precise measurement of noise and equipment limitations. Problems can also arise when parts exhibiting extreme output mismatch are to be characterized. Accurate characterization of such parts may require tuning at the output port as well as at the input.

A second and more accurate method measures the device noise figure at multiple source admittances, allowing an indirect characterization of the device noise parameters. This requires measurement of noise at a minimum of four unique source impedances. Measurements at additional source impedance levels, however, allow use of error correction methods to improve accuracy. For these reasons, more than four (typically nine or more) source admittances are preferable.

Extraction of device noise parameters is accomplished by considering equation (4.94), which relates device noise to the applied source admittance. Thus, given four known source admittances (Y_1, Y_2, Y_3, and Y_4) and the measured noise figure at each source impedance (F_1, F_2, F_3, and F_4), equation (4.94) allows an algebraic solution for device noise parameters F_{min}, R_{opt}, X_{opt}, and R_n.

When measurements at more than four admittances are performed, the accuracy of the extracted parameters is enhanced by using methods that minimize the error [34]. For example, a least-squares method minimizes the weighted square sum of the i'th measured data, which are given by

$$e = \left| F_{min} + \frac{R_n}{G_{gi}} (G_{gi} - G_{opt})^2 + (B_{gi} - B_{opt})^2 - F_i \right| \tag{4.95}$$

Additionally, measurement uncertainty also exists in defining the source admittance Y_{gi}. Techniques have been developed that consider minimizing this error [35].

4.4.2 Equivalent Circuit Elements with Noise Spectral Density

The empirical noise model proposed by Gupta (Chapter 2) characterizes the device with one noise measurement at the device drain-source terminals. A small-signal equivalent circuit model of the device at the bias levels of interest is also needed. In deriving this noise model, the equivalent device model is represented by five elements, including input resistance $R_T = R_g + R_s + R_i$, transconductance g_m, output resistance r_{ds}, and a white noise current source of spectral density S_{i0}. The first four circuit element values may be determined by using S-parameter measurements and the small-signal analysis described in Section 4.2. The white current source is characterized by the measurement methods described in Chapter 3.

REFERENCES

[1] M. Weiss and D. Pavlidis, "Power Optimization of GaAs Implanted FET's Based on Large-Signal Modeling," *IEEE Trans. Microwave Theory Tech.*, Vol. MTT-35, February 1987, pp. 175–188.

[2] M.A. Smith, T.S. Howard, K.J. Anderson, and A.M. Pavio, "RF Nonlinear Device Characterization Yields Improved Modeling Accuracy," *IEEE Microwave Theory Tech. Symp. Digest*, 1986, pp. 381–384.

[3] F. Diamand and M. Laviron, "Measurement of the Extrinsic Series Elements of a Microwave MESFET Under Zero Current Conditions," *Proc. 12th European Microwave Conf.*, Finland, September 1982, pp. 451–456.

[4] H. Kondoh, "An Accurate FET Modelling from Measured S-Parameters," *IEEE Microwave Theory Tech. Symp. Digest*, June 1986, pp. 377–380.

[5] E.W. Strid, "Extracting More Accurate FET Equivalent Circuits," *Monolithic Technol.*, October 1987, pp. 3–7.

[6] G. Dambrine, A. Cappy, F. Heliodore, and E. Playez, "A New Method for Determining the FET Small-Signal Equivalent Circuit," *IEEE Trans. Microwave Theory Tech.*, Vol. MTT-36, July 1988, pp. 1151–1159.

[7] T. Chen and M. Kumar, "Novel GaAs FET Modeling Technique for MMICs," *IEEE GaAs IC Symp. Digest*, 1988, pp. 49–52.

[8] W.R. Curtice and R.L. Camisa, "Self-Consistent GaAs FET Models for Amplifier Design and Device Diagnostics," *IEEE Trans. Microwave Theory Tech.*, Vol. MTT-32, December 1984, pp. 1573–1578.

[9] H. Fukui, "Determination of the Basic Device Parameters of a GaAs MESFET," *Bell Syst. Tech. J.*, Vol. 58, March 1979, pp. 771–797.

[10] W.R. Frensley, "Power-Limiting Breakdown Effects in GaAs MESFET's," *IEEE Trans. Electron Devices*, Vol. ED-28, August 1981, pp. 962–970.

[11]   S.H. Wemple, M.L. Steinberger, and W.O. Schlosser, "Relationship Between Power Added Efficiency and Gate-Drain Avalanche in GaAs MESFETs," *Electron. Lett.,* Vol. 16, June 5, 1980, pp. 459–460.

[12]   R. Vaitkus, "Uncertainty of the Values of the GaAs MESFET Equivalent Circuit Elements Extracted from Measured Two-Port Scattering Parameters," *Proc. IEEE/Cornell Conf. High-Speed Semiconductor Devices and Circuits,* August 15–17, 1983, pp. 301–308.

[13]   R.A. Minasian, "Simplified GaAs MESFET Model to 10 GHz," *Electron. Lett.,* Vol. 13, No. 8, 1977, pp. 549–551.

[14]   E. Arnold, M. Golio, M. Miller, and B. Beckwith, "Direct Extraction of GaAs MESFET Intrinsic Element and Parasitic Inductance Values," *IEEE Microwave Theory Tech. Symp. Digest,* May 1990, pp. 359–362.

[15]   B. Hughes and P.J. Tasker, "Bias Dependence of MODFET Intrinsic Model Element Values at Microwave Frequencies," *IEEE Trans. Electron Devices,* Vol. ED-36, October 1989, pp. 2267–2273.

[16]   M. Sango, O. Pitzalis, L. Lerner, C. McGuire, P. Wang, and W. Childs, "A GaAs MESFET Large-Signal Model for Nonlinear Analysis," *IEEE Microwave Theory Tech. Symp. Digest,* May 1988, pp. 1053–1056.

[17]   P. Ladbrooke, *MMIC Design: GaAs FETs and HEMTs,* Norwood, MA: Artech House, 1989.

[18]   W. Press, B. Flannery, S. Teukolsky, and W. Vetterling, *Numerical Recipes,* Cambridge, UK: 1989.

[19]   M. Miller, M. Golio, B. Beckwith, E. Arnold, D. Halchin, S. Ageno, and S. Dorn, "Choosing an Optimum Large Signal Model for GaAs MESFETs and HEMTs," *IEEE Microwave Theory Tech. Symp. Digest,* May 1990, pp. 1279–1282.

[20]   J. Bandler, Q.J. Zhang, and S.H. Chen, "Efficient Large-Signal FET Parameter Extraction Using Harmonics," *IEEE Trans. Microwave Theory Tech.,* Vol. MTT-37, December 1989, pp. 2099–2108.

[21]   D. Luenberger, *Linear and Nonlinear Programming,* Reading, PA: Addison-Wesley, 1984.

[22]   L.E. Scales, *Introduction to Non-linear Optimization,* London: Macmillan, 1985.

[23]   B. Flannery, S. Teukolsky, and W. Vetterling, *Numerical Recipes in C,* New Rochelle, NY: Cambridge University Press, 1988.

[24]   T.R. Cuthbert, Jr., *Optimization Using Personal Computers,* New York: John Wiley and Sons, 1987.

[25]   R. Fletcher, *Practical Methods of Optimization,* New York: John Wiley and Sons, 1987.

[26]   K. Levenberg, "A Method for the Solution of Certain Problems in Least Squares," *Quart. Appl. Math.,* Vol. 2, 1944, pp. 164–168.

[27]   D. Marquardt, "An Algorithm for Least-Squares Estimation of Nonlinear Parameters," *SIAM J. Appl. Math.,* Vol. 11, 1963, pp. 431–441.

[28]   S. Kirkpatrick, C.D. Gelatt, and M.P. Vecchi, "Optimization by Simulated Annealing," *Science,* Vol. 220, No. 4598, May 1983, pp. 671–680.

[29]   M. Vai, S. Prasad, and N.C. Li, "Modeling of Microwave Semiconductor Devices Using Simulated Annealing Optimization," *IEEE Trans. Electron Devices,* Vol. ED-36, No. 4, April 1989, pp. 761–762.

[30]   B.R. Epstein, S. Perlow, D. Rhodes, J. Schepps, M. Ettenberg, and R. Barton, "Large-Signal MESFET Characterization Using Harmonic Balance," *IEEE Microwave Theory Tech. Symp. Digest,* 1988, pp. 1045–1048.

[31]   H. Willing, C. Rauscher, and P. de Santis, "A Technique for Predicting Large-Signal Performance of a GaAs MESFET," *IEEE Trans. Microwave Theory Tech.,* Vol. MTT-26, December 1978, pp. 1017–1023.

[32]   J.M. Golio, R.J. Trew, G.N. Maracas, and H. Lefevre, "A Modeling Technique for Characterizing Ion-Implanted Material Using C-V and DLTS Data," *Solid-State Electron.,* Vol. 27, April 1984, pp. 367–373.

[33]   W.R. Curtice and S. Pak, "On-Wafer Verification of a Large-Signal MESFET Model," *IEEE Trans. Microwave Theory Tech.,* Vol. MTT-37, November 1989, pp. 1809–1811.

[34]   R.Q. Lane, "The Determination of Device Noise Parameters," *Proc. IEEE,* August 1969, pp. 1461–1462.

[35]   M. Mitama and H. Katoh, "An Improved Computational Method for Noise Parameter Measurement," *IEEE Trans. Microwave Theory Tech.,* Vol. MTT-27, June 1979, pp. 612–615.

Chapter 5
APPLICATIONS AND LIMITATIONS

The models, techniques, and procedures discussed in the previous chapters can be used to advantage in solving many of the design problems encountered by microwave circuit and device designers. When incorporated into circuit simulation routines, accurate device models allow engineers to design and optimize circuit performance prior to fabrication. Standing alone, device models often provide the information required when choosing an appropriate device for a specific application or when making decisions concerning device processing. The combination of physically based models and monolithic circuit technology potentially allows designers to perform simultaneous device and circuit optimization during the design process.

In this chapter, a number of microwave circuit applications that can be addressed by using MESFETs and HEMTs are presented. The discussions outline some of the performance parameters of interest for each application. In Chapter 2, we described a number of device models. The roles that these models play in the circuit design process are presented as part of our discussions.

An important task facing both circuit designers and device fabrication engineers is evaluating the ultimate potential of a device for specific applications. This information can be important for choosing an appropriate device for particular applications. Although such an evaluation can be accomplished by developing a model for the device and optimizing the performance using a circuit simulation routine, such an approach is usually prohibitively time-consuming. Instead, the problem is often first approached using easily derived performance figures of merit (FOMs). Many FOMs can be estimated directly from device models. Some of the more important FOMs as well as limitations in applying them are examined in Section 5.2.

Improvements in reported MESFET and HEMT performance have been made as each year passes. These improvements are arrived at primarily by refining the fabrication processes. More sophisticated processing techniques allow smaller devices to be manufactured. Improvements in performance by reductions in device size, however, are limited. These limits are imposed both by physical and techno-

---

logical constraints. Although advances in fabrication tools and procedures eventually erase the technological barriers, improvements in device performance are inevitably halted by the physical limitations of the device. When this occurs, the microwave community is forced to explore other devices to fill its needs. In the final sections of this chapter, issues related to the ultimate performance limits of MESFETs and HEMTs are examined.

5.1 APPLICATIONS

5.1.1 Low Noise and Small-Signal Amplifiers

Small-signal amplifiers are used throughout communication and EW systems. These circuits are designed to operate as linear gain blocks, with specific requirements determined by their location within the system. Most such systems must extract and process signals that are embedded in background noise. Any noise contributed by the system, therefore, must be kept low enough to ensure that the desired signals are not masked. To keep the overall system noise figure small, low noise amplifiers (LNAs) are used in the earliest stages of these electrical systems. The LNAs must have sufficiently low noise figures to meet system needs. In addition, they must also have adequate gain to ensure that the succeeding stages have little effect on the system noise figure. Small-signal amplifiers are also used as intermediate gain stages for RF or IF signals. In addition, they are often used as buffer amplifiers to isolate circuits from the poor voltage standing wave ratios (VSWRs) commonly encountered with mixers and oscillators, or to minimize load-pull effects on oscillators. In these applications, the noise figure of the amplifiers is usually less critical than the intermodulation distortion performance.

Before a significant amount of time is committed to the design of any amplifier, the proper device must be selected. The device S-parameters may be used in conjunction with noise parameters to assess the potential of a particular MESFET or HEMT's ability to meet the gain and noise performance requirements. The device parameters of most importance in this determination are the minimum noise figure, F_{min}, and the associated gain G_{as}. The noise parameters may be obtained for the device as described in Chapters 3 and 4, along with the models described in Chapter 2. The minimum noise figure of the device F_{min} may then be calculated at the desired frequencies, and G_{as} can be determined using the noise parameters and S-parameters in a linear simulator.

For cases in which intermodulation distortion concerns drive the amplifier design, large-signal simulations may be used to determine the suitability of a device. This is discussed in more detail in Section 5.1.2.

Although single-ended amplifiers are utilized for many applications, these types of LNAs typically exhibit poor input VSWRs due to the fact that the optimum noise impedance Z_{opt} is different from the optimum power matching imped-ance S_{11}^*. One method commonly used to improve the input VSWR is to use series feedback as illustrated in Figure 5.1. By adjusting the feedback and the load imped-ance, the difference between Z_{opt} and S_{11}^*, may be reduced, thereby improving the VSWR.

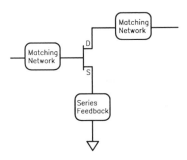

Figure 5.1 Simplified schematic of a common source amplifier showing the location of the series feed-back elements.

Another approach to improve the input VSWR is to use a balanced configu-ration, as shown in Figure 5.2. In this configuration, two identical amplifiers are connected between a pair of quadrature hybrids. Reflected energy from the ampli-fier inputs will be directed to the load resistor R_L, rather than to the input port, resulting in a low VSWR. HEMT-balanced amplifiers have demonstrated better than a 20-dB return loss across the 8.5- to 16-GHz frequency range with a gain and a noise figure of 9.7 and 3.5 dB, respectively [1].

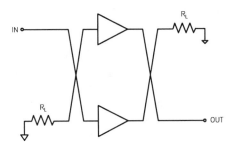

Figure 5.2 Balanced amplifier configuration.

Parallel feedback, shown in Figure 5.3, is often used in small-signal amplifier designs. A series resistor and inductor in the feedback path help to reduce the effects of frequency roll-off of the transistor by allowing greater feedback at lower frequencies and decreasing the feedback as frequency is increased. An additional benefit is that the input impedance to the transistor-feedback combination is reduced, making the input matching task simpler. The addition of the parallel elements, unfortunately, will degrade the noise figure, and thus may not be useful in all applications.

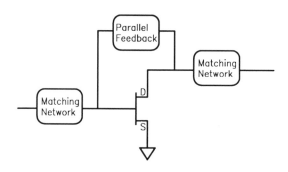

Figure 5.3 Parallel feedback in a common source amplifier.

Other topologies have also been used to develop LNAs. Both MESFET [2–5] and HEMT [6, 7] technologies have been used with great success in various broadband, low noise, and high frequency applications.

Small-signal amplifiers are typically operated at input power levels well below power saturation; thus, the design and analysis of the amplifier may be accomplished using small-signal S-parameters and noise parameters. While, for most designs, the use of measured S-parameters is ideal, small-signal equivalent circuits are required in many cases. For example, the designer may need to design an amplifier to work over a frequency range not covered by the measured S-parameter data. In this case, the small-signal equivalent circuit developed from the measured data may be used to extrapolate or interpolate the measured data to the frequency of interest. This approach has been used to develop many millimeter-wave amplifiers. Amplifiers at frequencies as high as 44 GHz with a 14.4-dB gain and a 3.6-dB noise figure have been developed using this extrapolation technique [8]. The device model in this case is obtained using measured S-parameter data to 25 GHz.

Standard small-signal equivalent circuit models also often simplify the development of feedback and matching circuitry for all of the above-mentioned design problems. For applications in which the noise figure is critical, noise models are also of great value. In the case of MMIC circuitry, the equivalent circuit and noise models offer the advantage of allowing the designer to consider gate width scaling as part of the design. This can be a very important part of the development of matching circuitry—especially for broadband applications.

5.1.2 Power Amplifiers

Solid-state power amplifiers are typically used in communication systems to provide sufficient signal power to allow transmission from one site to another. The popularity of mobile and hand-held communication has spurred the development of compact and efficient power amplifiers. Additionally, the size, weight, and power constraints on satellite communication systems require such amplifiers.

Power amplifiers are distinguished from small-signal amplifiers in that they are usually designed for maximum efficiency, rather than linearity or low noise. Output power and gain are also important parameters in the design of power amplifiers. The power amplifier design problem is largely a task of determining optimum trade-offs for these various performance characteristics.

Choosing an optimal device for a power amplifier application is often a difficult task. Given the large-signal model of a power device, the optimal input and output impedances for best power transfer may be determined using the model in conjunction with a nonlinear simulator. The power-added efficiency, gain, and intermodulation distortion levels of the device can also be determined from the simulation results. The process of determining the model is easily accomplished by applying the characterization and parameter extraction techniques described in Chapters 3 and 4. When the resources are available, load-pull measurements can also be used in the device selection and qualification process.

Many microwave MESFET power amplifiers operate in class A mode. In this case, the transistor is biased such that it is always "on." The theoretical maximum efficiency of this type of amplifier is 50%, though typical values realized in practice are 20 to 30%. Class A MESFET power amplifiers have been reported with output powers of 17 W and 32% power-added efficiency in the C band [9], and 2 W with 11% power-added efficiency in the Ka band [10]. Use of HEMTs is often reserved for low noise applications, though HEMT power amplifiers at 44 GHz with 250 mW of output power and 27% efficiency have been developed [11].

Class B amplifiers have theoretical maximum efficiencies of 78.5%. In this bias configuration, the transistors are biased near pinch-off, such that the quiescent current is less than a few percent of the saturation current. Current flows through the device only during the positive portion of the output waveform. Filtering of the output signal is required to remove the harmonics generated in this process, thereby restricting the useful bandwidth. Class C amplifiers are similar to class B, except that they are biased well into the pinched-off region, so current flows for even a smaller portion of the signal. In this case, even higher efficiencies are possible. Compared to class A amplifiers, lower power densities are achieved in class B and C configurations. Amplifiers biased for class B operation have been reported with about 2 W of output power and power-added efficiencies of 70% in the C band [9] and 45% at the X band [12].

Devices biased for class B operation may also be used in a push-pull configuration, reducing the output filtering requirements and thus extending the useful bandwidth. Often the transistors are biased for class AB operation to improve the intermodulation distortion characteristics. Using the push-pull configuration, a 6- to 15-GHz power amplifier has produced 1.85 W with an IP_3 of $+45$ dBm [13]. Although this performance is very good, many factors other than amplifier topology (device breakdown voltage, channel properties, *et cetera*) contribute to third-order intercept performance. Quantification of the improvement derived from utilizing class AB circuit topologies, therefore, is difficult.

One method used to improve the efficiency of narrow band class B amplifiers is to employ harmonic tuning in the output stages. In this topology, commonly referred to as class F, the even harmonics are terminated by a short and the odd harmonics are terminated with an open. A 450-mW 10-GHz amplifier has been developed using this approach, with a power-added efficiency of 61% using a single 1200-μm MESFET [14].

To obtain high output power levels, the total gate periphery of the active devices must be increased. This results in device impedances that are more difficult to match over wide frequency ranges. One approach to develop broadband power amplifiers is to use power-combining techniques as illustrated in Figure 5.4. In this case, the signal is divided among several paths, amplified through smaller active devices, and then recombined to obtain high output power levels. The total gate periphery is large, but each of the smaller devices may be impedance matched separately, simplifying the broadband matching problem. One disadvantage of this configuration is that the additional circuit losses and phase variations may degrade

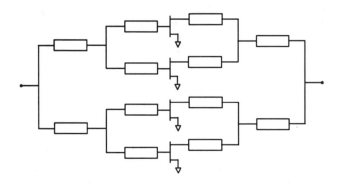

Figure 5.4 Power-combining amplifier.

the amplifier efficiency significantly. Power combining has been used [15] to develop a 25-W amplifier with 40% power-added efficiency at 5 GHz.

Both large- and small-signal models are used at various points in the power amplifier design process. Small-signal models are often used to develop the initial circuit designs. Although such models neglect important nonlinear characteristics of the device, they may be used to develop efficient first-order estimates of circuit component values. To determine final circuit element values for these applications, large-signal models are required. These models are used to determine output power, optimum device bias, and intermodulation distortion for the design.

Intermodulation distortion is often quantified by the third-order or second-order intercept points, IP_3 and IP_2, respectively. In low power applications, the device should have the best intermodulation characteristics when the output is matched for best power transfer, that is, with a load impedance of S_{22}^*. For devices that are driven harder, the optimal output impedance will not be S_{22}^*; it must be determined.

If desired, load-pull simulations may be performed to determine the optimal output match and gain of the device at the desired frequencies and output power levels. Load contours can be obtained by optimizing to specific power levels, though a large number of simulations are required to realize the contours.

When the optimal termination impedances are known, the matching networks can be designed using small-signal methods. Verification of the design then requires nonlinear analysis. Small-signal analysis using measured S-parameters or equivalent circuit models may be used to determine the small-signal stability of the amplifier, though this will not guarantee stability under large-signal conditions. Additional confidence may be obtained by using a large-signal model with a time-domain simulation to decrease computation time.

5.1.3 Oscillators

Oscillators are used in many systems, along with mixers, to perform frequency conversion of desired signals. Early solid-state microwave oscillators often used transferred electron devices. MESFET- and HEMT-based oscillators, however, have become more common. One key to their popularity is their high dc-to-RF conversion efficiencies. In addition, the three terminal devices allow greater design flexibility to meet specific performance requirements.

Many systems require extremely low noise oscillators, which cannot be realized using free running MESFET or HEMT oscillators. One method of improving the oscillator phase noise is to use dielectric resonators for stabilization. Some typical circuit configurations for dielectric resonator oscillators (DROs) are shown in Figure 5.5. Dielectric resonators are narrow bandwidth structures, limiting DRO usage to fixed LO applications.

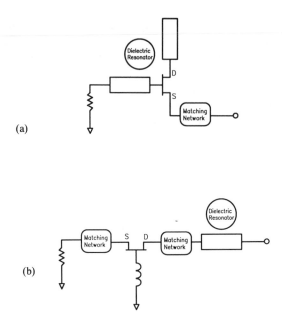

(a)

(b)

Figure 5.5 Common dielectric resonator oscillator topologies: (a) shunt feedback; (b) reflection oscillator.

In applications for which broadband frequency coverage is required, variable frequency oscillators are required. Two popular types are varactor-tuned and YIG-tuned oscillators. In the first case, varactor diodes are used as variable capacitors to adjust the resonant frequency of the oscillator. A sample configuration is shown in Figure 5.6. In a YIG oscillator, the magnetic field around a YIG sphere is altered

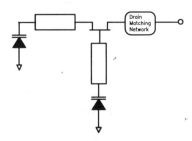

Figure 5.6 Typical topology for varactor-tuned oscillators.

to change the resonant frequency of the structure. At resonance, energy is reflected back into the active device, resulting in oscillations. A commonly used circuit configuration for a YIG oscillator is illustrated in Figure 5.7.

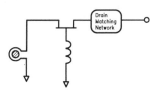

Figure 5.7 YIG oscillator topology.

The varactor-tuned oscillator is useful in applications requiring high tuning speed, moderate tuning range, and low power consumption. YIG oscillators achieve greater tuning range, with good noise performance; however, the inductance of the coupling loop of the YIG assembly limits the tuning speed. Additionally, the conversion efficiency of YIG oscillators is low, due to the high dc bias requirements. The YIG circuit requires high bias to drive the magnetic coils used to tune the oscillator. YIG oscillators are often used in precision test equipment, such as spectrum analyzers and sweep generators, as well as in electronic countermeasure systems.

A commonly used approach to oscillator design is to use *S*-parameters and small-signal simulations to design and analyze the circuits [16, 17]. A disadvantage to this method is that it does not account for the shift in device impedance as it becomes oscillatory. More important, information concerning output power from the oscillator circuit is not obtained from small-signal simulations. The small-signal to large-signal shift in performance will change the oscillation frequency from that predicted in the small-signal analysis. For this reason, an oscillator design should also be simulated using large-signal models and a nonlinear time-domain simulator. Harmonic balance simulators are also theoretically capable of performing large-signal oscillator analysis. The large-signal simulation provides both a more accurate estimate of the frequency of oscillation and a prediction of the oscillator's output power.

In addition to predicting oscillation frequency, linear simulations may also be used to estimate such parameters as tuning range, tuning linearity, and frequency pulling. Large-signal simulations are useful for predicting the spectral purity, tuning speed, settling time, and modulation capabilities. In addition, frequency pushing of the oscillator may be estimated using nonlinear analysis.

5.1.4 Mixers

A mixer is a three-port device that functions to convert an input RF signal in conjunction with an LO signal to an intermediate signal termed the IF. The IF signal may be either the sum or difference between the LO and RF signals and is identified as up conversion and down conversion, respectively. Ideally, a mixer performs this frequency conversion with perfect fidelity and thus generates no intermodulation distortion (IMD). Other desirable characteristics include high isolation between all three ports and a low noise figure, with all of these characteristics obtained with minimal loss or preferably gain while performing frequency conversion.

Mixers are essential in all system applications that require frequency conversion. In receiver system applications, for example, a LNA followed by a mixer are typically placed near the antenna to convert an RF signal to a baseband frequency. In this application, system requirements such as sensitivity (i.e., minimum signal level detection specifications) define the conversion loss and IMD characteristics of the mixer.

Traditionally, mixers have been based on the nonlinear voltage-current response of Schottky diodes in order to perform frequency conversion. While such mixers exhibit many of the desired characteristics listed above, all diode-based mixers exhibit conversion loss, ranging from perhaps 3 dB for image enhanced mixers to 10 dB for mixers with high IMD performance. Also, the local oscillator power needed to drive the diodes to achieve efficient frequency conversion is dictated by the desired IMD performance. Typical LO power requirements range from 0 to +20 dBm.

More recently, MESFETs and HEMTs are being used as the nonlinear element in mixers—primarily for reasons of compatibility with MMIC processing and to achieve conversion gain. Frequency conversion of the RF signal to an IF results in a larger amplitude signal. Due to conversion gain, rather than loss, the requirement for a LNA preceding the mixer may be eliminated in certain applications. This may also result in a net decrease in dc power, which is particularly important for portable and space-based equipment.

The FET-based mixers are generally categorized into one of three topologies: (1) gate, (2) drain, and (3) resistive mixers. An example of each topology is shown in Figure 5.8. For the gate mixer, both LO and RF signals are applied to the gate of the FET, while the IF is extracted from the drain terminal [18–20]. The MESFET or HEMT is biased near pinch-off such that the applied LO signal varies the FET transconductance over a highly nonlinear region. Frequency conversion occurs primarily due to the nonlinear FET transconductance and thus a gate mixer is sometimes referred to as a transconductance mixer. This mixer also requires filtering at the IF port to terminate reactively LO, RF, and intermodulation components as well as "diplex" the IF signal. For this reason, this mixer is best suited for applications in which the IF signal is substantially above or below the RF-LO fre-

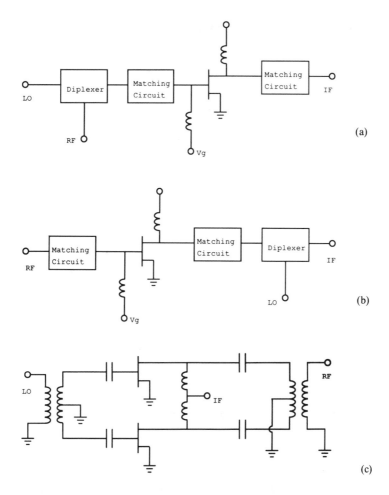

Figure 5.8 FET-based mixers are generally categorized into one of three types: (a) gate mixer; (b) drain mixer; (c) resistive mixer.

quency to allow such filtering. Based on HEMTs, such a mixer has demonstrated performance up to 94 GHz RF with approximately 6-dB conversion gain [21].

For the drain mixer, the LO signal is applied at the drain terminal, RF at the gate terminal, and the IF is extracted from the drain. Similar to the gate mixer, this mixer requires filtering to diplex the IF from the LO. The mixer operates with the FET drain-source voltage set near the knee of the linear and saturated regions of the I-V curve with a gate source bias of less than 0 V (Figure 5.9). At this bias condition, and with an LO signal applied to the drain, the FET transconductance and output resistance are both nonlinear and contribute to mixing. Compared to the gate topology, some improvement in noise performance has been reported with the drain topology [18].

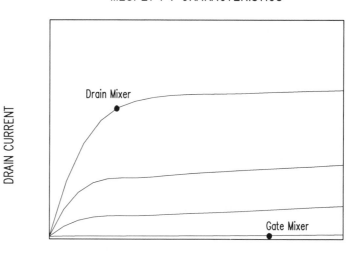

Figure 5.9 Gate and drain mixers are biased near pinch-off and the knee of the current-voltage curve, respectively.

In a resistive mixer, the FET is utilized as a time-varying resistor, which is modulated by the LO signal. No external bias voltages are applied to the FET. Under this bias condition, the FET is weakly nonlinear, and thus very high IMD performance is possible. However, because the FET is unbiased, and is operated as a time-varying resistor, it exhibits significant conversion loss. Third-order input intercept points of +30 dBm with about 8-dB conversion loss have been shown [22].

The design and analysis of FET-based mixers requires a means of predicting both dc and RF characteristics. To this end, the large-signal MESFET and HEMT models described in Chapter 2 are particularly important. Such models, when used in time-domain (i.e., SPICE) or harmonic simulators, allow predictions of dc as well as RF characteristics to be obtained. These simulators predict the spectral content at any node within the circuit. This information is useful in designing and analyzing the reactive terminations as well as predicting many of the mixer performance characteristics such as conversion loss, port-to-port isolation, return loss, and intermodulation products. Nearly all mixer performance characteristics, except for noise performance, can be predicted with the large-signal models and an appropriate simulator.

Because FET mixers are naturally nonlinear circuits that require the FET to operate over large gate and drain voltage fluctuations, possibly extending into breakdown, the large-signal FET models must represent the device over a large operating range. Extracting model parameters for this range naturally determine the model's accuracy. Hence, some loss of accuracy in predicting mixer circuit performance is expected. Some accuracy improvement can be obtained by using two large-signal models: One to predict primarily dc characteristics and a second one for the RF performance.

5.1.5 Attenuators

A variable attenuator is a two-port device that allows adjustment of the signal amplitude by application of an external voltage or current. The attenuation is proportional to voltage or current, and in cases where it is directly proportional, the attenuator is termed *linearized.* Attenuators are further categorized into step (digital) and continuously variable. Step attenuators exhibit discrete values in response to a digital stimulus, while continuously variable attenuators are controlled by an analog stimulus.

Ideally, an attenuator performs this function with perfect fidelity and thus generates no harmonic distortion. Other desirable characteristics include operation over a specified input power range, minimal insertion loss in the low attenuation state, a desired high attenuation range, a low I/O VSWR, and constant insertion loss with frequency.

System applications for attenuators are numerous. They are commonly employed in feedback networks, communication systems, and in temperature compensation networks to maintain constant signal amplitude. Such circuits find applications in systems for which adjustment of signal amplitude is required.

Topologies for variable attenuators are typically based on the conventional "pi" or "tee" circuits shown in Figure 5.10. These topologies consist of three resistors, including both series and shunt. Although attenuation and VSWR are dependent on all three resistor values, minimal attenuation is determined primarily by the smallest achievable value of series resistance. Another topology shown in Figure 5.10, which is termed the "bridge tee," exhibits variable attenuation by adjusting only two resistor values instead of three. In all cases, correct selection of resistance values results in the desired attenuation with near-perfect VSWR [23].

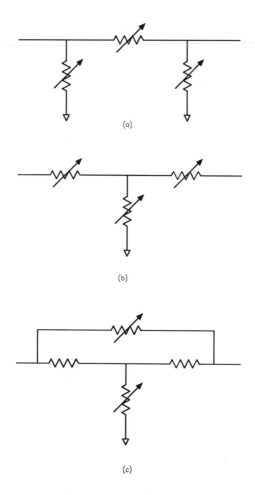

Figure 5.10 Most FET attenuators are based on traditional fixed attenuator topologies: (a) Pi attenuator; (b) Tee attenuator; (c) bridge Tee attenuator.

Traditionally, attenuators have been implemented with either series or shunt PIN diodes and a pair of couplers configured in a balanced topology. The PIN diode exhibits a resistance proportional to the applied dc current. Also, because PIN diodes exhibit low parasitic capacitance, they are highly insensitive to changes in frequency. PIN diodes, however, require considerable dc bias current to obtain low resistances and also require external circuitry for biasing.

More recently, MESFETs and HEMTs are being used to realize the variable resistive element in attenuators (Figure 5.11). In this case, the FET is unbiased ($V_{ds} = 0$, $I_{ds} = 0$), and the drain-source resistance r_{ds} is varied with an applied gate potential. Thus, very little dc bias current is needed to change the drain-source resistance r_{ds}. Unlike a PIN-based attenuator, FET-based circuits are voltage controlled and require very little dc power.

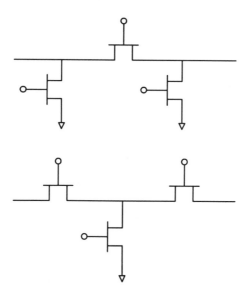

Figure 5.11 In FET attenuators, the drain-source resistance r_{ds} is varied by changing the gate potential.

FET-based attenuators have demonstrated performance beyond 40 GHz with a minimal insertion loss of 2.6 dB (at 40 GHz) and a 30-dB dynamic attenuation range [24]. Typical attenuator designs, however, are less broadband, with the FET physical dimensions or attenuator topology chosen for a particular bandwidth to optimize performance [25–27].

The FET equivalent circuit models discussed in Chapter 2 provide considerable insight into the operation and design of attenuators. Since the FET is unbiased ($V_{ds} = 0$, $I_{ds} = 0$), the transconductance is zero, and this results in the simplified equivalent FET model shown in Figure 5.12. While the attenuator operation relies primarily on varying r_{ds} with gate potential, other elements such as the capacitances and parasitic resistances are also important. These elements greatly influence frequency effects, as well as minimum and maximum attenuation. The FET scaling principles discussed in Chapter 2 allow evaluation of trade-offs in selecting device gate width. For example, increasing the gate width of the device results in a smaller minimum value of r_{ds} (and thus less loss), but capacitances C_{ds}, C_{gs}, and C_{gd} become larger. These larger capacitances obviously limit the dynamic range at high frequencies.

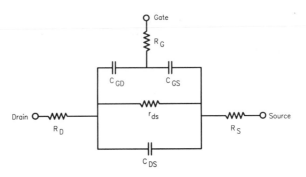

Figure 5.12 Equivalent circuit of a FET for an attenuator circuit.

The design and analysis of FET-based attenuators require a means of predicting both dc and RF characteristics. The large-signal MESFET and HEMT models described in Chapter 2 are particularly important. When used in time-domain (i.e., SPICE) or harmonic simulators, predictions of both dc and RF characteristics can be obtained. These simulators predict the spectral content at any node within the circuit. This information is useful in designing and predicting many characteristics such as attenuation, return loss, and intermodulation products—all as a function of gate bias. Nearly all performance characteristics can be predicted with the large-signal models and an appropriate simulator.

Because attenuators require the FET to operate over large gate voltages and with zero drain-source voltage, the large-signal FET models must represent the device over a considerable range. Extraction of the model parameters must include both dc and RF characteristics. Because all large-signal models presented in Chapter 2 cannot simultaneously match both RF and dc characteristics, some loss of accuracy in predicting the dc gate control and attenuation performance is expected.

5.2 FIGURES OF MERIT

The device models presented in Chapter 2 can be used in conjunction with the characterization and parameter extraction techniques of Chapters 3 and 4 to define the performance characteristics of a MESFET or HEMT. The resulting model is used with circuit simulation software to predict a wide range of performance characteristics. Making preliminary judgments, however, about the ultimate performance potential of devices or about which devices should be chosen for a particular circuit design application is often desirable. Similarly, determination of where to optimally bias a device to achieve specific electrical behavior is sometimes desirable. In these cases, we prefer to arrive at precursory decisions without having to design and simulate a circuit completely.

The first-order calculation of several performance FOMs can be very useful for making preliminary judgments concerning active device capabilities. The information obtained from these FOMs should be useful for making foundry decisions, choosing discrete devices, or evaluating the performance potential of newly developed devices.

The starting point for estimation of many FOMs is either microwave characterization data or determination of a complete equivalent circuit model for the device. The techniques described in Chapters 3 and 4 should be used to accomplish these tasks. FOMs of particular interest include gain-bandwidth product f_T; maximum frequency of oscillation f_{max}; maximum stable gain, MSG; maximum gain efficiency, E_G; minimum noise figure, F_{min}; and associated gain, G_{as}.

Characterization data and FOMs do not completely describe all behavior that must be considered in choosing an appropriate foundry or device for a given application. Considerations such as cost, delivery, reliability, and yield, for example, are not addressed by these data. Technical issues other than device behavior may be more important for some applications than the performance of active devices. For example, the ability to produce RF via holes or high quality mixer diodes during MMIC fabrication may be of greater concern than the performance of the MESFET or HEMT when choosing a foundry. Likewise, the technical FOMs are only first-order indicators of ultimate performance limits. The data obtained in this fashion, however, can be used as a basis of coarse comparisons of active devices.

5.2.1 Gain-Bandwidth Product

The gain-bandwidth product of a MESFET or HEMT is the frequency at which the short-circuit current gain of the device falls to unity—or, in terms of h-parameters, the frequency at which

$$|h_{21}| = 1 \tag{5.1}$$

By using the simplified equivalent circuit shown in Figure 5.13, the magnitude of the short-circuit gain of the MESFET or HEMT is easily computed to be

$$G_I = \frac{g_m}{\omega C_{gs}} \tag{5.2}$$

Setting this quantity to unity and solving for the appropriate frequency yields

$$f_T = \frac{g_m}{2\pi C_{gs}} \tag{5.3}$$

Although equation (5.3) is often utilized to predict the gain-bandwidth product, the value computed using this formula will dramatically overestimate actual device performance in many applications. This is especially true for devices with sub-half-micron gate lengths. The inaccuracy of the equation is due primarily to the use of the highly simplified equivalent circuit of Figure 5.13, which neglects the effect that gate-drain capacitance has on device behavior. A more accurate expression for the gain-bandwidth product can be derived from circuit analysis of a more complete equivalent circuit model that includes gate-drain capacitance. Such analysis leads to the expression:

$$f_T = \frac{g_m}{2\pi(C_{gs} + C_{gd})} \tag{5.4}$$

In practice, for modern microwave MESFETs and HEMTs, the f_T actually measured using the $|h_{21}| = 1$ definition is typically on the order of 5 to 10% lower than that calculated by using equation (5.4).

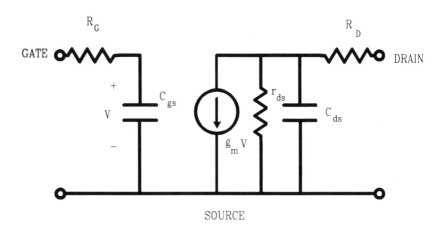

Figure 5.13 Simplified small-signal equivalent circuit model of the MESFET or HEMT used to obtain equation (5.2).

Note that the gain-bandwidth product of a device is bias dependent. Figures 5.14 and 5.15 present the computed f_T values for a microwave MESFET and HEMT device, respectively. The devices used to produce Figures 5.14 and 5.15 will be used throughout Section 5.2 to compare typical MESFET and HEMT FOM characteristics.

Figures 5.14 and 5.15 show not only that the f_T values for the devices depend on bias, but also that the dependence is very different for a MESFET than for a

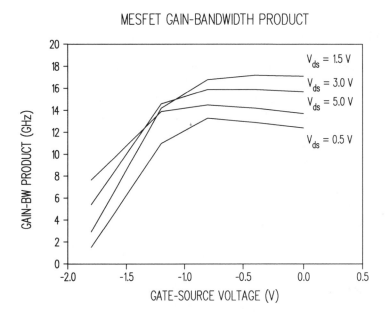

Figure 5.14 Gain-bandwidth product of a 0.5×300-μm MESFET as a function of bias.

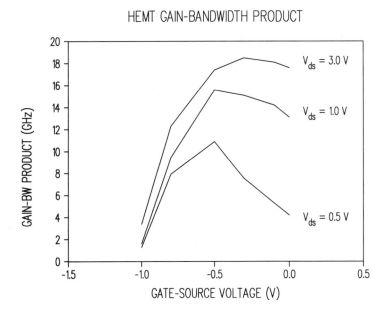

Figure 5.15 Gain-bandwidth product of a 0.7×200-μm HEMT as a function of bias.

HEMT. The reasons for these differences are apparent on examination of equation (5.4). Both device transconductance and gate-source capacitance play key roles in determination of the gain-bandwidth product of the device. The bias dependence of these characteristics is shown in Sections 1.3 and 1.5 to be quite different for the MESFET than for the HEMT.

To first order, the gain-bandwidth product of a device is independent of gate width. This is consistent with expressions (5.3) and (5.4) because g_m, C_{gd}, and C_{gs} values are all directly proportional to device gate width. In contrast, f_T approximates the inverse of gate length because transconductance increases while device capacitances decrease with reduced gate length. Note that the gain-bandwidth product for the HEMT presented in Figure 5.15 is comparable to the MESFET gain-bandwidth product of Figure 5.14, although the HEMT gate length is 40% greater than the MESFET gate length. This trend of improved high frequency performance for HEMTs over MESFETs is typical of modern devices.

The FOM f_T does not represent the limiting frequency of operation for microwave circuitry. Power gain may still be obtained from the device beyond the f_T frequency. This FOM can, however, be used to compare the approximate frequency performance limits of different devices. In general, the device with a high f_T value will function usefully at higher frequencies than a device with a much lower f_T value.

5.2.2 Maximum Frequency of Oscillation

The maximum frequency of oscillation f_{max} is the highest frequency at which power gain can be obtained from a device. This FOM, like f_T, may be used as an indicator of the ultimate frequency limits of a device. As with the f_T value, a high value of f_{max} is desirable if high frequency operation is of interest.

For most microwave applications, the frequency f_{max} appears to be a more useful FOM than the frequency f_T because microwave designers are typically concerned with power gain into conjugately matched conditions (not short-circuit current gain). Unfortunately, the maximum frequency of oscillation typically occurs well above the measurement capabilities of characterization equipment commonly available in microwave labs.

The maximum frequency of oscillation is defined formally as the frequency at which the unilateral power gain of a device goes to unity:

$$|U| = 1 \tag{5.5}$$

where U can be written in terms of device y-parameters [28] to be

$$U = \frac{|y_{21} - y_{12}|^2}{4(g_{11}g_{22} + g_{12}g_{21})} \tag{5.6}$$

and where $g_{ij} = \text{Re}[y_{ij}]$.

For microwave MESFETs and HEMTs being produced in research laboratories today, the frequency at which equation (5.5) is satisfied can be well in excess of 100 GHz. This makes it extremely difficult to measure f_{max} directly. Extrapolation of data measured at 18 or 26 GHz to frequencies this high is often invalid [29, 30].

An alternative to the direct measurement of the maximum frequency of oscillation is to estimate the value from analysis of the device equivalent circuit. This method is also susceptible to error, but can be used to obtain first-order estimates of a device's high frequency performance potential. As with gain-bandwidth product, more than one approximate expression for f_{max} can be obtained from evaluation of equivalent circuits for the device. A first-order expression that is often used to determine the maximum frequency of oscillation of the device can be expressed as [31]:

$$f_{max} = \frac{f_T}{2} \left[\frac{r_{ds}}{R_{gt}} \right]^{1/2} \tag{5.7}$$

where r_{ds} is the device output resistance, R_{gt} is the sum of the gate parasitic resistance R_g and the charging resistance R_i, and f_T is the gain-bandwidth product for the device.

An expression that more accurately approximates f_{max} and takes into account the gate-drain capacitance effect is given by [16, 32]:

$$f_{max} = \frac{f_T}{2} (r_1 + f_T^* T_3)^{-1/2} \tag{5.8}$$

where

$$r_1 = \frac{R_g + R_i + R_s}{r_{ds}}$$

and

$$T_3 = 2\pi R_g C_{gd}$$

As in the case of gain-bandwidth product, the maximum frequency of oscillation of a device depends on the bias conditions. Figures 5.16 and 5.17 present f_{max} values

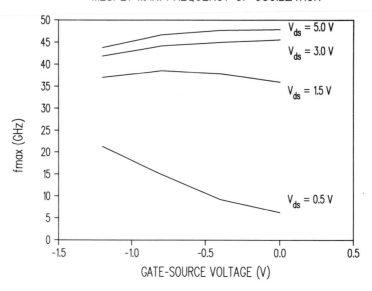

Figure 5.16 Maximum frequency of oscillation of a 0.5×300-μm MESFET as a function of bias.

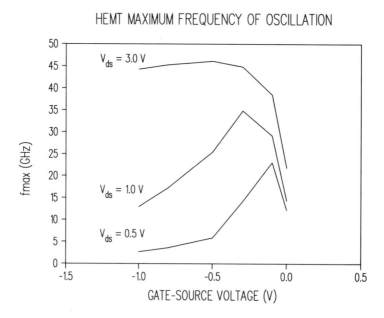

Figure 5.17 Maximum frequency of oscillation of a 0.7×200-μm HEMT as a function of bias.

computed for the MESFET and HEMT discussed earlier using equation (5.8). Note the qualitative differences in the MESFET and HEMT curves. These are expected because of the relationship between f_{max} and f_T expressed by equations (5.7) and (5.8).

Although neither f_T nor f_{max} is an ideal measure of the ultimate frequency capabilities of a device, when both figures are considered in a comparison, some insight is gained into the high frequency performance of devices relative to one another.

5.2.3 Maximum Stable Gain

A number of different power gain definitions can be used for the purposes of comparing microwave MESFETs and HEMTs. Maximum available gain (MAG), maximum stable gain (MSG), and unilateral power gain (U), are all commonly discussed in the literature [28–33]. Each of these gain definitions can provide useful information concerning device performance potential. A relationship between the various gain definitions can also be derived [29, 30].

MSG is a very useful FOM because it is defined under all bias conditions and is easily determined from measured S-parameter data. This is in contrast to maximum available gain, which is defined only when the device is unconditionally stable. The MSG represents the highest possible value of power gain that is achieved before instability occurs. The formula for MSG is

$$MSG = \frac{|S_{21}|}{|S_{12}|} \tag{5.9}$$

Clearly, a high value of MSG over the frequency range of interest for a given application is desirable. Most commercially available small-signal circuit simulation routines are capable of performing MSG calculations and displaying them in standard output formats.

The value of MSG is not only bias dependent, but also depends on frequency. Although maximum stable gain provides better specific information for circuit design purposes than either f_T or f_{max}, more information is required to use MSG for device performance evaluation.

Figures 5.18 and 5.19 present measured MSG for the MESFET and HEMT, respectively. The information is presented for the two devices biased at both I_{dss} and $\frac{1}{2}I_{dss}$. Characteristics of both the MESFET and HEMT are similar, with maximum stable gain falling at a rate of 3 dB per octave. The MSG begins to fall at 6 dB per octave at higher frequencies, and at even higher rates near the frequency f_{max}.

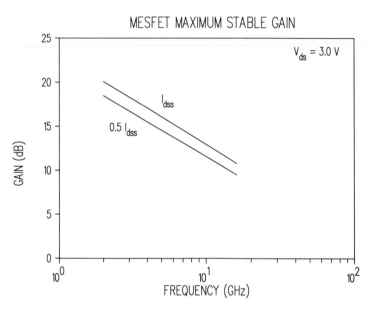

Figure 5.18 Maximum stable gain of a 0.5×300-μm MESFET as a function of frequency for two different bias levels.

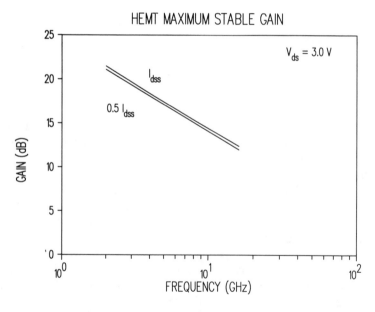

Figure 5.19 Maximum stable gain of a 0.7×200-μm HEMT as a function of frequency for two different bias levels.

5.2.4   Maximum Gain Efficiency

For many microwave applications, dc power consumption is a more critical design constraint than the gain of a device. As a first-order measure of maximum gain efficiency, E_G can be defined:

$$E_G = \frac{\text{MSG}}{I_{ds}} \tag{5.10}$$

Figure 5.20 presents measured maximum gain efficiency as a function of frequency for the microwave MESFET utilized in this section. Although the MSG is seen from Figure 5.18 to be higher when the device is biased at I_{dss}, the gain efficiency is significantly higher when biased at half this current level. This trend is also observed for the HEMT device. These results are illustrated in Figure 5.21. Note that although the HEMT has very comparable f_T, f_{max}, and MSG values to the MESFET, its gain efficiency is considerably higher. This is typical of modern HEMT devices.

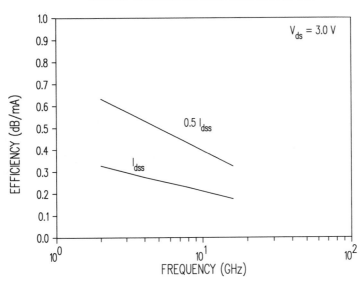

Figure 5.20 Maximum gain efficiency of a 0.5×300-μm MESFET as a function of frequency for two different bias levels.

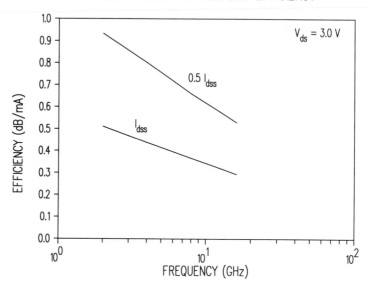

Figure 5.21 Maximum gain efficiency of a 0.7×200-μm HEMT as a function of frequency for two different bias levels.

5.2.5 Minimum Noise Figure and Associated Gain

The minimum noise figure F_{min} represents the minimum achievable noise figure of a device at a given bias level when operated with the input circuit matched for ideal noise performance. In general, this ideal noise match is not identical to the ideal match for power gain applications. The power gain that the device is capable of achieving when matched for optimum noise performance is the associated gain G_{as}. This pair of parameters clearly must be considered together when low noise design is to be accomplished. A low value of noise figure and a high value of associated gain are desirable for LNA applications. These parameters are discussed in detail in Sections 3.4 and 4.4.

5.3 MESFET LIMITATIONS

The performance of microwave MESFETs improves as each year passes. Useful gain is obtained from MESFETs at higher and higher frequencies. Minimum noise figures continue to be reduced and higher maximum output power is obtained from

MESFETs. Most of these improvements in device capabilities are accomplished by the reduction of the size of the device (scaling). However, physical and practical limits exist to how small devices can be made while still exhibiting useful microwave performance properties.

Scaling relationships that lead to optimal microwave devices are not easily determined. When one dimension is reduced, other dimensions in the device must also be reduced, but not necessarily in a linear fashion. Likewise, if proper performance is to be obtained, doping concentrations must be increased when the device size is reduced. Although physically based models and simulations of devices can be helpful in this process, device design generally depends considerably on empirical data and prior experience.

One way to investigate the MESFET scaling problem is to examine device dimensions, doping densities, and performance figures of previously fabricated devices. Although the data collected in this fashion may be scattered, definite trends emerge. From the data, common relationships between device dimensions and active layer doping densities are discovered. These relationships can be expressed mathematically and extrapolated to very small device dimensions. By considering the effects certain physical mechanisms and practical constraints have on scaling, the ultimate limits on device size can be predicted. Finally, performance figures can be extrapolated to these devices of minimum size to predict ultimate performance limits for devices.

5.3.1 MESFET Scaling Rules and Physical Limits

Examination of a large number of MESFETs fabricated and reported in the literature since 1966 has produced three different scaling rules that can be used to describe the collected data [34]. Each of the three rules is determined by considering different aspects of the collected data.

Rule 1 is established by comparing the published data to three different previously proposed scaling relationships. Those relationships include:

(1) constant LN_d product scaling [35],
(2) constant L^2N_d product scaling [36], and
(3) constant $L^{1/2}N_d$ product scaling [35].

Of the proposed relationships, constant LN_d scaling best describes the devices presented in published reports [34]. The accumulated data are approximated by the rule 1 relationship, which is expressed as

$$N_d = \frac{1.6 \times 10^{17}}{L} \tag{5.11}$$

where gate length L is given in microns and doping density is in cm^{-3}.

Rule 2 is determined by computing a least-squares straight-line fit to reported doping density and gate length data [34]. No consideration of physical principles is applied to obtain this rule. The resulting expression is

$$N_d = \frac{1.5 \times 10^{17}}{L^{1.43}} \tag{5.12}$$

where again gate length and doping density are given in microns and cm^{-3}, respectively.

Finally, rule 3 is developed by considering not only doping density and gate length, but also the effect doping density has on electron transport (i.e., low field electron mobility). The relationship between electron mobility and doping density is presented in Section 1.1. Inclusion of the transport properties of the active channel in the scaling rule results in the expression:

$$N_d = \frac{6.0 \times 10^{16}}{\mu_0 L} [(V\text{-}s)^{-1}] \tag{5.13}$$

where μ_0 is the electron mobility in the active channel. In equation (5.13), units of mobility, doping concentration, and gate length must be consistent with each other.

The doping density-gate length scaling rules of equations (5.11) through (5.13) are plotted in Figure 5.22 along with the collected MESFET data. The statistics plotted in Figure 5.22 include all of the GaAs MESFET data of this earlier work along with additional reported MESFET characteristics. The references used to accumulate this information are listed in Table 5.1. These device reports are used to produce all of the high frequency and low noise data presented in this section.

We cannot conclude that any of the three outlined scaling relationships will produce an *optimal* device because each of the rules is determined empirically. We can reasonably assume, however, that any of the rules will produce a good first estimate of optimal scaling relationships. By utilizing any of the three rules, the required channel doping density for a device can be estimated from the device gate length.

In addition to the channel doping concentration levels, device epi-layer thickness is also scaled with gate length in a GaAs MESFET. In general, the thickness of the epitaxial layer is reduced with decreasing gate length. Several physical and practical constraints, however, determine the range of acceptable epi-thickness values for a given gate length or channel doping density. These considerations limit the ultimate dimensions that a MESFET can possess while still exhibiting acceptable performance.

To achieve improved frequency performance from a scaled device, the gate length to epi-thickness ratio of the device needs to be greater than about π. This scaling relationship can be derived analytically by considering the effect the L/a

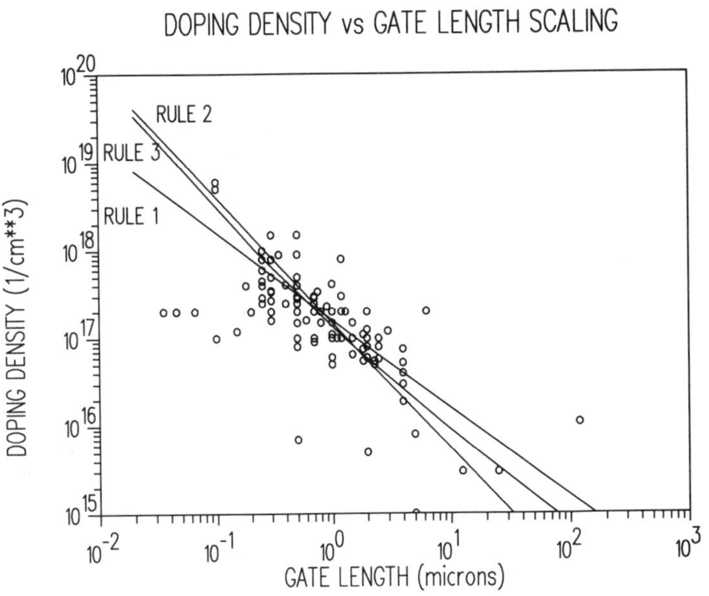

Figure 5.22 Reported gate length and doping density plotted along with the scaling rules given by equations (5.11) through (5.13). The devices used to produce the figure are from the reports of Table 5.1.

Table 5.1
Published Reports of GaAs High Frequency and Low Noise MESFETs Fabricated Since 1966

1. Mead, *Proc. IEEE*, February 1966
2. Hooper, *Proc. IEEE*, July 1967
3. Waldner, *Proc. IEEE*, November 1969
4. Shapiro, *Proc. IEEE*, November 1969
5. Statz, *Solid-State Electron.*, December 1969
6. Drangeid, *Electron. Lett.*, April 1970
7. Lehovec, *Solid-State Electron.*, October 1970
8. Hower, *Semiconductors and Semimetals*, Academic Press, 1971
9. Driver, *Proc. IEEE*, August 1971
10. Baechtold, *IEEE Trans. Electron Devices*, May 1972
11. Jutzi, *IEEE Trans. Electron Devices*, March 1972
12. Hower, *IEEE Trans. Electron Devices*, March 1973
13. Sitch, *Proc. IEEE*, March 1973
14. Hunsperger, *Electron. Lett.*, June 1973
15. Baechtold, *Electron. Lett.*, March 1973
16. Dean, *IEEE Microwave Theory Tech.*, December 1973
17. Statz, *IEEE Trans. Electron Devices*, September 1974
18. Liechti, *IEEE Microwave Theory Tech.*, May 1974
19. Anastassiou, *IEEE Microwave Theory Tech.*, February 1974
20. Fair, *IEEE Trans. Electron Devices*, June 1974
21. Hunsperger, *Solid-State Electron.*, May 1975
22. Maeda, *Proc. IEEE*, February 1975
23. Maeda, *IEEE Trans. Electron Devices*, August 1975

Table 5.1 continued

24. Barrera, *IEEE Trans. Electron Devices,* November 1975
25. Liechti, *IEEE Microwave Theory Tech.,* June 1976
26. Baudet, *IEEE Microwave Theory Tech.,* June 1976
27. Pucel, *IEEE Microwave Theory Tech.,* June 1976
28. Hornbuckle, *IEEE Microwave Theory Tech.,* June 1976
29. Irie, *IEEE Microwave Theory Tech.,* June 1976
30. Ogawa, *IEEE Microwave Theory Tech.,* June 1976
31. Barnes, *IEEE Trans. Electron Devices,* September 1976
32. Ohata, *IEEE Trans. Electron Devices,* August 1977
33. Kellner, *Solid-State Electron.,* May 1977
34. Kohn, *Solid-State Electron.,* January 1977
35. Mimura, *Proc. IEEE,* September 1977
36. Abe, *IEEE Microwave Theory Tech.,* March 1978
37. Sone, *IEEE Trans. Electron Devices,* March 1978
38. D'Arso, *IEEE Trans. Electron Devices,* October 1978
39. Takahashi, *IEEE Trans. Electron Devices,* October 1978
40. Furutsuka, *IEEE Trans. Electron Devices,* June 1978
41. Suzuki, *IEEE Microwave Theory Tech.,* December 1979
42. Fukui, *IEEE Microwave Theory Tech.,* July 1979
43. Hashizume, *IEEE Trans. Electron Devices,* March 1979
44. Fukui, *IEEE Trans. Electron Devices,* July 1979
45. Mizuishi, *IEEE Trans. Electron Devices,* July 1979
46. Fukui, *Solid-State Electron.,* May 1979
47. deSantis, *IEEE Microwave Theory Tech.,* June 1980
48. Niclas, *IEEE Microwave Theory Tech.,* March 1980
49. Ohata, *IEEE Trans. Electron Devices,* June 1980
50. Weinreb, *IEEE Microwave Theory Tech.,* October 1980
51. Suyama, *IEEE Trans. Electron Devices,* June 1980
52. Imai, *IEEE Electron Devices Lett.,* March 1981
53. Levy, *IEEE IEDM,* December 1981
54. Yokoyama, *IEEE IEDM,* December 1981
55. Dekkers, *IEEE Trans.Electron Devices,* September 1981
56. Tajima, *IEEE Trans. Electron Devices,* February 1981
57. Oxley, *IEEE IEDM,* December 1981
58. Beneking, *IEEE Trans. Electron Devices,* May 1982
59. Morcok, *IEEE Trans. Electron Devices,* February 1982
60. Feng, *IEEE Electron Devices Lett.,* November 1982
61. Chye, *IEEE Electron Devices Lett.,* December 1982
62. Yokoyama, *IEEE Trans. Electron Devices,* October 1982
63. Chen, *IEEE Trans. Electron Devices,* October 1982
64. Feng, *IEEE Trans. Electron Devices,* November 1983
65. Gupta, *IEEE Trans. Electron Devices,* December 1983
66. Chao, *IEEE Electron Devices Lett.,* September 1983
67. Bandy, *IEEE Electron Devices Lett.,* February 1983
68. Imai, *IEEE Electron Devices Lett.,* April 1983

Table 5.1 continued

69. Chao, *IEEE Electron Devices Lett.*, April 1983
70. Kohn, *IEEE Electron Devices Lett.*, April 1983
71. deSalles, *IEEE Microwave Theory Tech.*, October 1983
72. Gupta, *IEEE Microwave Theory Tech.*, December 1983
73. Ohta, *IEEE Trans. Electron Devices*, March 1984
74. Tomizawa, *IEEE Trans. Electron Devices*, April 1984
75. Daembkes, *IEEE Trans. Electron Devices*, August 1984
76. McNally, *IEEE Electron Devices Lett.*, April 1984
77. Feng, *IEEE Electron Devices Lett.*, March 1984
78. Feng, *IEEE Electron Devices Lett.*, January 1984
79. Mondal, *IEEE Microwave Theory Tech.*, October 1984
80. Ayasli, *IEEE Microwave Theory Tech.*, January 1984
81. Niclas, *IEEE Microwave Theory Tech.*, March 1984
82. Chao, *IEEE Trans. Electron Devices*, June 1985
83. Ishida, *IEEE Trans. Electron Devices*, June 1985
84. Bastida, *IEEE Trans. Electron Devices*, December 1985
85. Watkins, *IEEE Microwave Theory Tech. Symp.*, June 1985
86. Nair, *IEEE Trans. Electron Devices*, September 1986
87. Fischer, *IEEE Trans. Electron Devices*, February 1986
88. Gutmann, *IEEE Microwave Theory Tech. Symp.*, June 1986
89. Gupta, *IEEE Electron Devices Lett.*, October 1987
90. Zeghbroeck, *IEEE Electron Devices Lett.*, May 1987
91. Hughes, *IEEE Trans. Electron Devices*, April 1987
92. Mishra, *IEEE*/Cornell, 1987
93. Zhou, *IEEE Microwave Theory Tech. Symp.*, 1987
94. Hung, IEEE Microwave Theory Tech., December 1988
95. Bandla, *IEEE Microwave Theory Tech. Symp.*, 1988
96. Hegazi, *IEEE Microwave Theory Tech. Symp.*, 1988
97. Schindler, *IEEE Microwave Millimeter-Wave Monolithic Circuits*, 1988
98. Gupta, *IEEE Microwave Theory Tech.*, April 1988
99. Lan, *IEEE Microwave Millimeter-Wave Monolithic Circuits*, 1988
100. Enoki, *IEEE Trans. Electron Devices*, June 1988
101. Banerjee, *IEEE Electron Devices Lett.*, January 1988
102. Pao, *IEEE Electron Devices Lett.*, March 1988
103. Harder, *IEEE Electron Devices Lett.*, April 1988
104. Lo, *IEEE Electron Devices Lett.*, August 1988
105. Onodera, *IEEE Electron Devices Lett.*, August 1988
106. Tehrani, *IEEE Trans. Electron Devices*, May 1988
107. Bernstein, *IEEE Trans. Electron Devices*, July 1988
108. D'Avanzo, *IEEE GaAsIC-S*, 1988
109. Chang, *IEEE Trans. Electron Devices*, October 1988

ratio has on the important frequency FOMs, f_T and f_{max} [34]. Similar conclusions can also be obtained using other analysis techniques [37–39]. The condition places a constraint on allowable epi-thickness, which can be expressed as

$$a < \frac{L}{\pi} \tag{5.14}$$

For very small devices, a second physical constraint—breakdown voltage—can limit the maximum allowable epi-thickness in a GaAs MESFET. The breakdown voltage of a device with fixed dimensions decreases with increased doping. Breakdown effects are discussed in Section 1.1.5. This phenomenon is well documented as a fundamental limiting mechanism in GaAs power devices [40–42]. Although the breakdown phenomenon is complex, Frensley [40] has developed a simple semiempirical expression for device breakdown voltage as a function of channel doping density and epi-layer thickness. The expression relates observed gate-drain breakdown voltage to channel doping and epi-thickness and can be written as

$$a = \frac{4.4 \times 10^{13} \ (V/cm^2)}{N_d V_{br}} \tag{5.15}$$

where V_{br} is the gate-drain breakdown potential.

Low breakdown voltage values limit the maximum bias and signal levels that can be used in a circuit utilizing the device. Such a constraint does not determine an absolute scaling limit, but indicates that useful applications of the device will be more difficult to find as breakdown voltages decline.

Assuming a minimal drain-source bias of approximately 3 V, an approximate 1-V peak-to-peak signal level, and a minimal gate bias level requires that device breakdown voltage be greater than about 5 V. This minimum acceptable breakdown voltage can be translated into a maximum acceptable epi-layer thickness, which is given by

$$a < \frac{8.8 \times 10^{12} \ (cm^{-2})}{N_d} \tag{5.16}$$

Equations (5.14) and (5.16) represent two conditions that must be maintained if effective device scaling is to be accomplished. Although equation (5.14) expresses epi-layer thickness in terms of gate length, while equation (5.16) expresses this quantity in terms of doping density, the two conditions can be used along with any of the scaling rules of equations (5.11) through (5.13) to obtain identical information.

Practical limits also constrain the minimum acceptable epi-layer thickness of a scaled MESFET. In particular, small epi-layer thickness leads to small device pinch-off voltage. When the gate potential becomes more negative than the pinch-off voltage, the device is cut off and no current flows. In contrast, when the device gate terminal is forward biased to a value on the order of or slightly lower than the built-in potential of the gate terminal Schottky contact, forward conduction occurs and device performance degrades rapidly. A maximum dynamic range of the device is limited, therefore, by the difference between the pinch-off and forward conduction voltage. Although the built-in potential of a Schottky gate on GaAs is nearly constant (approximately 0.8 V), pinch-off is determined by doping density and epi-layer thickness.

For a practical microwave device to have a useful dynamic range, the epi-layer thickness must be great enough to allow reasonable voltage-current swings without pinching off the device or causing large gate forward conduction currents to flow. This condition can be expressed in terms of the built-in potential and channel Debye length as [34]:

$$a > \left(\frac{2\epsilon V_{bi}}{qN_d}\right)^{1/2} + 6L_D$$

or expressing the Debye length, L_D, in terms of channel doping density:

$$a > \left(\frac{2\epsilon V_{bi}}{qN_d}\right)^{1/2} + 6\left(\frac{\epsilon kT}{q^2 N_d}\right) \tag{5.17}$$

Another barrier to continued device scaling is the onset of significant tunneling currents through the Schottky barrier gate electrode on highly doped material. When a Schottky contact is made on a GaAs surface, a depletion region is formed at the metal-semiconductor interface as described in Section 1.2. The extent of that depletion region is approximately proportional to the square root of the channel doping density. Thus, as devices are scaled, channel doping densities are increased and depletion region penetration is reduced. Once this penetration becomes shallow enough, significant tunneling through the contact takes place and the terminal becomes nearly ohmic. When this occurs, the device reliability is seriously degraded and it no longer operates like a MESFET. To avoid this condition, we must ensure that the channel doping concentration remains below a level of about 1×10^{19} cm^{-3} [34]. This is expressed as

$$N_d < 1.0 \times 10^{19} \text{ cm}^{-3} \tag{5.18}$$

When the maximum and minimum epi-layer thickness constraints of equations (5.14), (5.16) and (5.17) are considered along with scaling rules of (5.11) through (5.13), minimum gate length dimensions for optimal GaAs MESFETs can be estimated. Figure 5.23 plots the maximum and minimum allowable epi-layer thickness of a GaAs MESFET (using scaling rule 1) as a function of device gate length. The curve marked "a_{max}" is determined by equation (5.14). For this scaling rule, the breakdown voltage considerations expressed in equation (5.16) are never as limiting as the frequency scaling considerations of equation (5.14). The curve labeled "a_{min}" is determined by equation (5.17). The two curves are seen to cross at a gate length dimension of about 0.2 μm, making optimal scaling impossible for gate length dimensions below this value. The abrupt rise in minimum epi-layer thickness for a gate length of about 0.015 μm corresponds to the device dimension when channel doping density exceeds 1.0×10^{19} cm^{-3} according to scaling rule 1. Devices scaled to this dimension will violate equation (5.18) and so are not expected to operate properly.

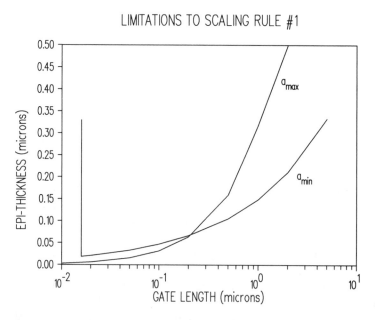

Figure 5.23 Maximum and minimum acceptable epi-thickness values determined for scaling rule 1.

Figures 5.24 and 5.25 plot corresponding data for scaling rules 2 and 3 expressed by equations (5.12) and (5.13). For these rules, maximum epi-layer thickness for long-gate-length devices is determined from equation (5.14). As the device is scaled to gate length dimensions in the 0.1- to 0.3-μm range, however, equation (5.16) becomes more limiting. As in the case of Figure 5.23, minimum epi-layer dimensions are determined from equation (5.17). Note that for scaling rules 2 and 3, device scaling is possible down to a gate length dimension of nearly 0.1 μm.

LIMITATIONS TO SCALING RULE #2

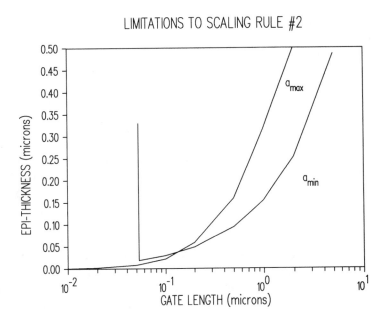

Figure 5.24 Maximum and minimum acceptable epi-thickness values determined for scaling rule 2.

LIMITATIONS TO SCALING RULE #3

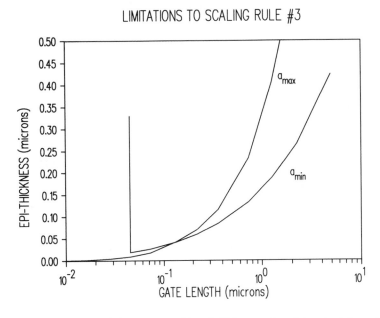

Figure 5.25 Maximum and minimum acceptable epi-thickness values determined for scaling rule 3.

The results presented above indicate that MESFETs that utilize gate dimensions much smaller than about 0.1 μm will not exhibit significantly improved performance. Although further reductions in device gate length are technologically possible, such devices will not exhibit improved frequency characteristics or will be useful only for a very limited number of applications.

5.3.2   Transconductance and Gate-Source Capacitance

Figures 5.26 and 5.27 present reported transconductance and gate-source capacitance as a function of gate length for GaAs MESFETs reported in the technical literature. The curves presented in these figures are obtained by computing the least-squares straight-line fit to the data.

The transconductance of the devices increases with decreasing gate length. This is the trend expected from first-order principles. The straight-line fit for transconductance as a function of gate length reaches a value of 338 mS/mm for a gate length dimension of 0.1 μm. The gate length dimension of 0.1 μm is chosen because this represents the scaling limit implied from the analysis of the previous subsection. The g_m value of 338 mS/mm lying on the least-squares fit line is obtained

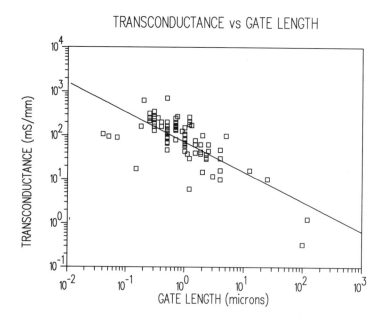

Figure 5.26 Reported transconductance per unit gate width as a function of reported MESFET gate length. The devices used to produce the figure are from the reports of Table 5.1.

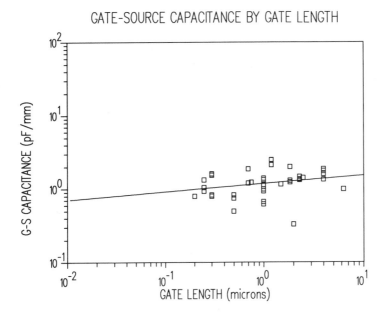

Figure 5.27 Reported gate-source capacitance per unit gate width as a function of reported MESFET gate length. The devices used to produce the figure are from the reports of Table 5.1.

using all available g_m data. Such a value underestimates the highest value of transconductance achievable because it makes use of data taken on devices that were not ultimate performers as well as those that approach optimum. The transconductance data from only the best devices reported in the technical literature can be used to argue for g_m values at a gate length of 0.1 μm that are much higher (exceeding 1000 mS/mm).

Note that the transconductance values plotted in Figure 5.26 consist of a combination of dc and RF values. As discussed in Chapter 1, however, measured RF transconductance is typically 5 to 25% lower than measured dc values. The fact that the plot is comprised of both types of data accounts for much of the observed scatter.

Gate-source capacitance data as a function of device gate length are presented in Figure 5.27. In contrast to the transconductance data, these characteristics exhibit very little gate length dependence. When any of the scaling rules specified by equations (5.11) through (5.13) are used, the reduction of device gate length requires increases in doping density. Although a reduction in gate length reduces gate-source capacitance, an increase in doping density increases it. These two scaling procedures tend to negate each other, and gate-source capacitance of properly scaled devices remains nearly constant at approximately 1 pF/mm of gate width.

5.3.3 Maximum Frequency of Oscillation and Gain-Bandwidth Product

Figure 5.28 presents maximum frequency of oscillation f_{max} as a function of MESFET gate length. As discussed in Section 5.2, values for f_{max} can be obtained either by extrapolating measured power gain data or by computing the maximum frequency of oscillation from available equivalent circuit element values using equation (5.7) or (5.8). Data plotted in Figure 5.28 are obtained primarily by computation of f_{max} from equivalent circuit information. When a published value was available, this information was used to estimate f_{max}.

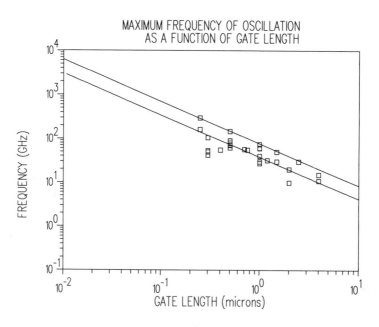

Figure 5.28 Reported or calculated maximum frequency of oscillation, f_{max}, as a function of reported MESFET gate length.

As can be seen in the figure, the least-squares fit line to all of the data (lower curve) predicts an ultimate f_{max} value of 341 GHz for a gate length of 0.1 μm. The equation for the f_{max} least-squares fit is given by

$$f_{max} = 38.05L^{-0.953} \qquad (5.19)$$

The upper curve in Figure 5.28 shows the predicted optimal f_{max} value as a function of gate length, which results if only the highest reported values for this FOM are used to extrapolate ultimate limits. A frequency of approximately 700 GHz is associated with a 0.1-μm-gate-length device for this case.

Gain-bandwidth product, f_T, as a function of gate length is presented in Figure 5.29. As in the case of maximum frequency of oscillation, this value can be obtained either by direct measurement of current gain as a function of frequency, or from equivalent circuit calculations. Furthermore, several different expressions for the gain-bandwidth product (all based on equivalent circuit calculations) are in common use. These expressions provide various degrees of accuracy. Reports of f_T values often do not indicate the method utilized to obtain this FOM. Many of the recently published values of f_T are determined from extrapolation of the $|h_{21}|$ *versus* frequency curve to unity. Some of the observed scatter in the data presented in Figure 5.29 is due to the differences in the methods used to compute f_T.

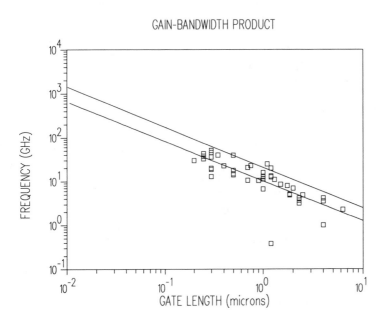

Figure 5.29 Reported or calculated gain-bandwidth product f_T as a function of reported MESFET gate length.

For convenience, the first-order expression given by equation (5.3) is used to compute f_T when reported values were not available. As noted in Section 5.2, this expression will overestimate the gain-bandwidth product of a device with small-gate-length dimensions.

In Figure 5.29, most of the f_T values plotted for devices with gate length dimensions of 0.25 μm or less lie on or below the least-squares fit line. Explanations that can be offered for this less than outstanding performance include (1) physical limitations are already being reached and FET scaling cannot continue to advance device performance or (2) the fabrication technology required to optimally process such devices has not yet matured. Based on the analysis of the previous section, higher f_T values for these small-gate-length devices will probably eventually be achieved. This will occur as fabrication procedures to deal with special problems of short-gate-length devices have been developed or improved. All three of the proposed scaling rules expressed in equations (5.11) through (5.13) suggest that a 0.25-μm-gate-length device be doped to a level of approximately 1×10^{18} cm^{-3} and have an epi-layer thickness of less than 0.08 μm. The status of processing technology today makes it difficult to achieve superior quality layers with these specifications.

Among the problems unique or more critical to the small-gate-length devices are transient velocity phenomena (for example, velocity overshoot), channel-substrate interface effects, and the effects of surface states. Each of these phenomena affects short-gate-length device behavior more significantly than long-gate-length device behavior. To scale MESFETs successfully and repeatably to sub-0.25-μm-gate-length dimensions, solutions to the problems posed by each of these phenomena have to be developed.

Velocity overshoot occurs in semiconductor devices when the electric field within the material changes abruptly over a very short distance or rapidly in time. As the doping density in the semiconductor material is increased, shorter dimensions are required for velocity overshoot to be observed. For the combination of channel dimensions and doping densities required for optimal MESFET scaling, the transient velocity phenomenon is not a dominant factor in the ultimate device performance. Although not of first-order importance, velocity overshoot may have a second-order effect on both the f_T and $f_{\max}$ performance. This phenomenon as it applies to the MESFET is discussed in Section 1.1. Monte Carlo simulation results are presented in that discussion. These simulations indicate that such phenomena might result in secondary performance deviations from those expected with only conventional transport mechanisms at work.

The occurrence of velocity overshoot in the active channel of a MESFET or HEMT can enhance or degrade device performance. Higher carrier velocity can be translated into higher frequency operation under certain circumstances, but can also mean less control (i.e., lower transconductance) of those carriers. The overall effect of velocity overshoot depends on many aspects of the device structure and the region of operation. Quantification of such effects for a given device is very difficult, even when multidimensional Monte Carlo device simulations are available.

Channel-substrate interface traps and surface traps can affect device behavior in very direct and important ways. Solving problems regarding these trapping states may require that more complex processing techniques or device structures be developed. Unwanted injection of channel carriers into the substrate must also be minimized to achieve optimum scaling. Reduction of the undesirable effects at the channel-substrate interface (traps and carrier injection into the substrate), for example, may require fabrication of the active channel on wide-band-gap buffer layers. This technique will result in much greater confinement of current carriers in the active GaAs channel. Accomplishing this confinement will reduce the number of carriers that interacts with substrate deep levels. A solution to the related problem of surface states may require fabrication of buried channel structures [43].

A least-squares straight-line fit to the data of Figure 5.29 is expressed by

$$f_T = 10.20L^{-0.905} \tag{5.20}$$

An f_T value of 82 GHz is predicted when equation (5.20) is evaluated for a gate length of $L = 0.1$ μm. Use of only the highest reported f_T data to extrapolate the ultimate frequency limits of a 0.1-μm device (indicated by the upper curve of Figure 5.29) produces a prediction that a gain-bandwidth value of nearly 200 GHz may be achievable from an optimally scaled GaAs MESFET.

Equation (5.20) can be compared directly to the expression derived by Fukui [44] for gain-bandwidth product as a function of gate length. The Fukui expression is

$$f_T = \frac{9.4}{L} \tag{5.21}$$

where, when L is given in microns, f_T is predicted in GHz. Figure 5.30 presents this comparison for devices with gate lengths between 0.05 and 5.0 μm. The two expressions are in remarkable agreement except at very small gate length dimensions.

5.3.4 Noise Figure Limits

Minimum noise characteristics as a function of gate length are more difficult to analyze than the previously presented FOMs. The measured minimum noise figure of a device is dependent on the frequency of the measurement. Thus, fewer data are available at any one frequency than are available for other performance FOMs. The gain associated with a particular noise figure performance is also of importance. A device with slightly higher noise figure but significantly more associated gain may be preferred for a given application. Scaling efforts, therefore, do not always address minimum noise figure performance as a primary goal. These issues make ultimate limit projections less accurate.

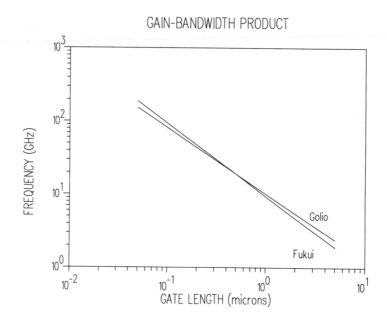

Figure 5.30 A comparison of the f_T values projected from the data collected in Table 5.1 to that projected by Fukui [44].

Figure 5.31 presents the measured minimum noise temperature at 8 GHz as a function of device gate length. The least-squares fit to the data projects a noise temperature of 39.5° for an optimally scaled 0.1-μm-gate-length device. This corresponds to a minimum noise figure of 0.54 dB. The minimum noise figure for a 0.1-μm device at 2, 4, 12, and 18 GHz is projected in an identical manner. The resulting data are plotted in Figure 5.32. Although the 12-GHz optimal noise figure prediction is low relative to the remaining data, the general trend illustrated in Figure 5.32 is as expected. Projected MESFET minimum noise figures throughout the X band are seen to be in the 0.3- to 0.6-dB range for an optimally scaled 0.1-μm device.

5.3.5 Output Power Limits

Power MESFET device scaling rules are not generally identical to, or even compatible with, scaling rules for optimal frequency and low noise performance. Because gate-drain breakdown is a primary limiting mechanism in power devices, increasing the breakdown voltage is a principal goal of the scaling. For equivalent gate lengths, slightly lower doping densities and larger epi-layer thickness values are

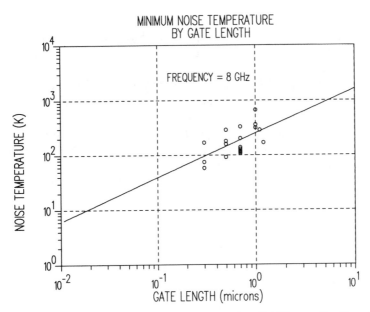

Figure 5.31 Reported minimum noise temperature measured at 8 GHz as a function of reported MESFET gate length.

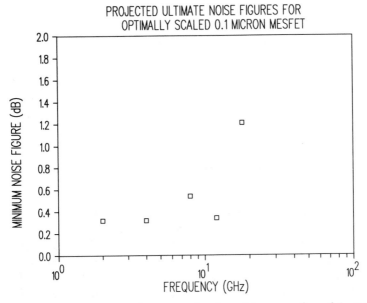

Figure 5.32 Projected minimum noise figure as a function of frequency for a 0.1-μm-gate-length MESFET.

typical of MESFETs used for power as opposed to low noise high frequency applications. Power devices also tend to utilize large gate widths compared to medium power or low noise devices.

Defining a single FOM that adequately summarizes device potential for power applications is difficult. Although maximum output power is clearly of great importance, the magnitude of the gain and the power-added efficiency is equally important in many circuits and systems. Figure 5.33 presents the relationship between these three FOMs for one particular power device [45]. Both the device gain and efficiency are seen to decrease with increasing output power. This trend is typical of devices operated near the maximum output power conditions.

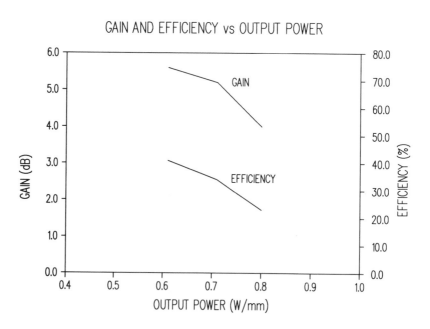

Figure 5.33 Gain and power-added efficiency as a function of output power for a GaAs MESFET.

Figure 5.34 presents maximum output power reported for several recently fabricated devices. Table 5.2 lists the reports used to prepare Figures 5.34 through 5.37. The data presented in Figure 5.34 comprise only the performance of devices with gate lengths in the 0.4- to 0.5-μm range. As the figure indicates, the devices are measured at frequencies that fall well below the maximum predicted f_{max} values for 0.5-μm devices. The output powers range from approximately 23 dBm at 35 GHz [46] to more than 44 dBm at 4.7 GHz [15]. A general trend of diminishing power with increasing frequency is also observed.

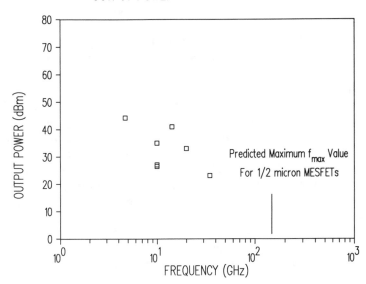

Figure 5.34 Reported maximum output power as a function of frequency for several recently fabricated power MESFETs with gate lengths in the 0.4- to 0.5-μm range.

Table 5.2
Recent Published Maximum Output Power Reports for GaAs MESFETs

1. Bechtle, *IEEE Trans. Microwave Theory Tech. Symp.*, 1987
2. Arasurala, *IEEE Trans. Microwave Theory Tech. Symp.*, 1988
3. Yamada, *IEEE Trans. Microwave Theory Tech. Symp.*, 1988
4. Geissberger, *IEEE Trans. Electron Devices*, May 1988
5. Tan, *IEEE Trans. Microwave Theory Tech.*, June 1988
6. Auricchio, *IEEE Trans. Microwave Theory Tech. Symp.*, 1989
7. Winslow, *IEEE Trans. Microwave Theory Tech. Symp.*, 1989
8. Kim, *IEEE Trans. Microwave Theory Tech.*, September 1989
9. Taniguchi, *IEEE Trans. Microwave Theory Tech. Symp.*, 1990
10. Khatibzedah, *IEEE Trans. Microwave Theory Tech. Symp.*, 1990

*Reported data are plotted in Figures 5.34 through 5.37.

A more commonly quoted FOM for power devices is the maximum power delivered per unit gate width. From first-order theoretical considerations, we can infer that the maximum output power obtained from a device is proportional to gate width. Such a relationship, however, has a limited range of validity. Large-gate-width devices provide greater dc levels with corresponding greater power-handling capabilities. This power-handling capability, however, is accomplished at the expense of increased device capacitance. High values of capacitance make delivery of power into and out of the device difficult at high frequencies and, thus, reduce the maximum usable frequency of the device.

Figures 5.35 and 5.36 present power per unit gate width and efficiency for the devices reported in Table 5.2. The power per unit gate width of modern power MESFETs is seen to fall between values of about 0.34 and 1.5 W/mm. The power-added efficiency numbers for the same devices fall in the range of 23 to 61%. At a given frequency, note that high power density values are associated with lower efficiency values.

The power gain of the power devices of Table 5.2 is plotted in Figure 5.37. The reported power gain is a generally decreasing function of frequency for power MESFETs. This is as expected from the results presented in Figure 5.32.

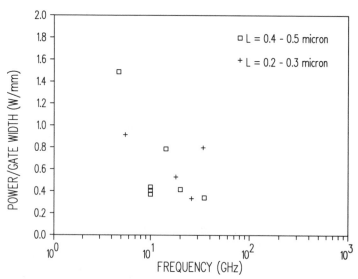

Figure 5.35 Reported output power per unit gate width as a function of frequency for the MESFETs listed in Table 5.2.

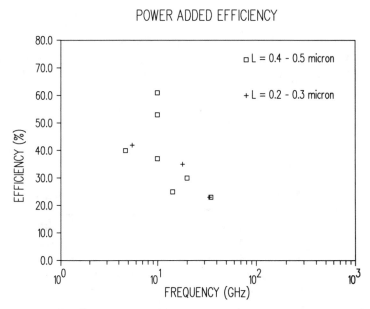

Figure 5.36 Reported power added efficiency as a function of frequency for the MESFETs listed in Table 5.2.

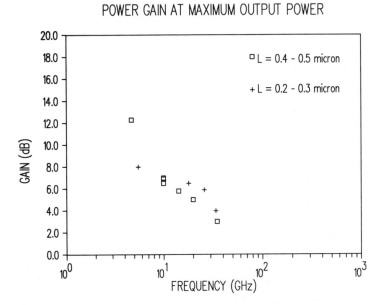

Figure 5.37 Reported power gain as a function of frequency for the MESFETs listed in Table 5.2.

Figures 5.32 through 5.37 illustrate the difficulties involved in the projection of ultimate power performance characteristics of microwave MESFETs. The data from recently fabricated devices do indicate, however, that power density values in excess of 1 W/mm should be achievable from devices at frequencies significantly below the ultimate f_{max} values predicted from Figure 5.28. Power-added efficiency values at these power levels are expected to be in the 40 to 60% range with associated gains of approximately 5 to 6 dB. Greater efficiency or gain values can be achieved at the expense of total output power.

5.4 HEMT LIMITATIONS

As in the case of the MESFET, performance characteristics reported for microwave and millimeter-wave HEMTs continue to improve. Minimum noise figures are reduced, while the frequencies of operation and maximum output powers are increased. Reducing device dimensions while increasing doping density levels (scaling) are important ingredients for obtaining these improvements.

Unlike MESFETs, HEMTs have been produced only since the mid-1980s. As a result, fewer reports of device geometries and associated microwave performance exist in the literature. The maturity of many HEMT fabrication processes also lags behind that available for the MESFET. Similar gaps exist in the simulation and modeling capabilities for the two devices. To further complicate investigations of HEMT limitations, the term "HEMT" applies to a broader class of device structures that is more complex than the MESFET (see Chapter 1).

Because of the relative immaturity of HEMT technology, analyzing HEMT scaling limits is difficult using the statistical-empirical techniques described in Section 5.3 and applied to the MESFET. We can, however, compare the relative effects of various limiting mechanisms in the MESFET and HEMT. Despite significant differences in the details of operation of the two devices, many of the physical mechanisms that will ultimately limit MESFET scaling will also limit HEMT scaling [47].

5.4.1 HEMT Scaling Rules and Physical Limits

Optimal scaling of the HEMT requires that the doping density of the AlGaAs layer lying directly beneath the gate be increased as the gate length is decreased. An expression for the optimal relationship between this doping density and gate length is difficult to define either from first-order models or the limited published information available. The optimum doping density for a given gate length will also depend on whether the HEMT is a conventional, pseudomorphic, InP-based, or multiple quantum-well structure. These structures are discussed briefly in Section 1.5. In general, the doping densities required for optimal HEMT scaling are higher than doping densities in equivalent gate length MESFETs [47].

In Section 5.3, several considerations that will ultimately limit the scaling of GaAs MESFETs were identified. Each of these considerations is quantified and expressed in equations (5.14) through (5.18). Analogous considerations also apply to HEMTs. Differences in the semiconductor structure and in the materials utilized in HEMTs alter the analysis slightly.

Equation (5.14) specifies the maximum epi-layer thickness that can be utilized for a given MESFET gate length if high frequency performance is to be improved from the scaling process. The equation is derived by considering the ratio of depletion charge stored under the gate electrode to that stored in fringing regions of the device [34]. The same equation holds for the HEMT if the epi-thickness of the MESFET, a, is replaced with the doped AlGaAs layer thickness of the HEMT, d (see Figures 1.21 and 1.44). The HEMT scaling condition to ensure improved frequency characteristics becomes

$$d < \frac{L}{\pi} \qquad (5.22)$$

The second limiting mechanism discussed for MESFET scaling, avalanche break-down, can also limit the scaling dimensions that can be realized in HEMTs. The breakdown voltage is a function of the semiconductor material in which the electric field lies. Calculated breakdown voltages for AlAs are approximately 30% higher than for GaAs [48]. This suggests that AlGaAs breakdown voltages will be 10 to 15% higher than GaAs material with equivalent ionized impurity concentrations. As mentioned previously, however, required doping density levels in HEMTs exceed those in MESFETs of equivalent gate length. This increased doping density lowers the observed breakdown voltage of the structure. In practice, this latter effect often dominates, making breakdown voltage considerations more limiting for HEMTs than for MESFETs. The resulting restriction reduces the maximum allow-able doped AlGaAs layer thickness for small-gate-length HEMTs. The general func-tional form for maximum epi-thickness to avoid breakdown limits in the HEMT gate-drain region is given by

$$d < \frac{K}{N_d} \qquad (5.23)$$

where K is expected to be slightly lower than the 8.8×10^{12} cm^{-2} value determined for GaAs MESFETs.

The third consideration for MESFET scaling limits is expressed in equation (5.17), but does not apply directly to the HEMT structure. The condition specifies a minimum acceptable epi-thickness to ensure the existence of a minimal con-ducting channel in the device. Because the conducting channel in the HEMT is not within the doped material, but exists in the quantum well formed at the heterostructure boundary, equation (5.17) does not apply. The HEMT has several advantages over the MESFET with respect to this minimal channel requirement. First, Debye length considerations do not apply to the HEMT analysis because the depletion region and active channel are separated. Second, the built-in potential of the Schottky barrier on a HEMT's AlGaAs layer is significantly greater than that of the GaAs Schottky barrier. This allows greater forward-bias signal levels to be applied without forward conduction taking place. Finally, the conduction band discontinuity of the HEMT structure acts to reduce the minimum required doped layer thickness. For the conventional HEMT structure, a condition for the minimum thickness of the doped AlGaAs layer can be obtained from consideration of equations (2.206) and (2.207). This condition can be expressed as

$$d_{min} = \left(\frac{2\epsilon V_{bi}}{qN_d} - \Delta E_c \right)^{1/2} \tag{5.24}$$

Note that this value of d_{min} will be significantly less than the value of a_{min} expressed in equation (5.17) for equivalent doping concentration levels.

The final mechanism discussed in Section 5.3, which could potentially limit MESFET scaling, is the onset of significant tunnel currents through the Schottky barrier. For Schottky barrier junctions on GaAs, this was seen to occur for doping densities on the order of 10^{19} cm^{-3}. This doping concentration level corresponds to device dimensions much smaller than those achievable without encountering other constraints. The increased built-in potential of a Schottky barrier on AlGaAs significantly increases this critical doping density level—thus reducing the potential for this phenomenon to become important.

In summary, constraints to HEMT scaling are similar to those for the MESFET. The maximum doped AlGaAs layer thickness for a given gate length HEMT will be limited according to equation (5.22) for long-gate-length dimensions, but by breakdown considerations and equation (5.23) at shorter gate lengths. The breakdown phenomenon will cause the maximum acceptable AlGaAs layer thickness to be smaller than an epi-layer thickness for a MESFET of equivalent gate length. Constraints on minimum AlGaAs layer thickness, however, will be considerably less restricting than analogous constraints on MESFET epi-layer thickness. The ultimate scaling limits for HEMTs will therefore be similar to, or slightly smaller than, those discovered for the MESFET (i.e., an approximately 0.1-μm minimum gate length).

To estimate how these scaling limits will affect HEMT electrical performance, performance data have been collected and analyzed for a large number of reported HEMT devices. The references used to accumulate this information are listed in Table 5.3. These device reports are used to produce all of the high frequency and low noise data presented in this section.

Table 5.3

Published Reports of High Frequency and Low Noise HEMTs Fabricated Since 1985

1. Gupta, *Electron Devices Lett.*, February 1985
2. Sheng, *Electron Devices Lett.*, June 1985
3. Chao, *Electron Devices Lett.*, October 1985
4. Mishra, *Electron Devices Lett.*, March 1985
5. Smith, *Electron Devices Lett.*, February 1985
6. Rosenberg, *Electron Devices Lett.*, October 1985
7. Sovero, *Electron Devices Lett.*, March 1986
8. Saunier, *Electron Devices Lett.*, September 1986
9. Tanaka, *Trans. Microwave Theory Tech.*, December 1986
10. Jay, *Trans. Electron Devices*, May 1986
11. Maas, *Trans. Microwave Theory Tech.*, July 1986
12. Chao, *Electron Devices Lett.*, October 1987
13. Yau, *IEEE Microwave Theory Tech. Symp.*, V. 2 1987
14. Bandy, *Trans. Microwave Theory Tech.*, December 1987
15. Asai, *IEEE Microwave Theory Tech. Symp.*, V. 2 1987
16. Shibata, *IEEE Microwave Theory Tech. Symp.*, V. 2 1987
17. Mishra, *IEEE Microwave Theory Tech. Symp.*, V. 2 1987
18. Awano, *Electron Devices Lett.*, October 1987
19. Nishimoto, *IEEE Microwave Theory Tech. Symp.*, V. 1 1987
20. Hayashi, *IEEE Microwave Theory Tech. Symp.*, V. 2 1987
21. Upton, *IEEE Microwave Theory Tech. Symp.*, V. 2 1987
22. Smith, *IEEE Microwave Theory Tech. Symp.*, V. 2 1987
23. Chen, *Trans. Microwave Theory Tech.*, December 1987
24. Sasaki, *IEEE Microwave Theory Tech. Symp.*, 1988
25. Van Hove, *Electron Devices Lett.*, October 1988
26. Hikosaka, *Electron Devices Lett.*, May 1988
27. Chow, *IEEE Microwave Theory Tech. Symp.*, V. 2 1989
28. Fathimulla, *Electron Devices Lett.*, July 1988
29. Geok Ng, *Electron Devices Lett.*, September 1988
30. Yuen, *Trans. Microwave Theory Tech.*, December 1988
31. Ito, *IEEE Microwave Theory Tech. Symp.*, 1988
32. Nguyen, *Trans. Electron Devices*, May 1989
33. Ketterson, *Trans. Electron Devices*, October 1989
34. Brown, *Trans. Electron Devices*, April 1989
35. Geok Ng, *Trans. Electron Devices*, October 1989
36. Wang, *Electron Devices Lett.*, February 1989
37. Lee, *Trans. Microwave Theory Tech.*, December 1989
38. Huang, *Electron Devices Lett.*, November 1989
39. Mishra, *Trans. Microwave Theory Tech.*, September 1989
40. Brown, *Electron Devices Lett.*, December 1989
41. Kao, *Electron Devices Lett.*, December 1989
42. Saito, *IEEE Microwave Theory Tech. Symp.*, 1989
43. Khanna, *Electron Devices Lett.*, December 1989
44. Metze, *IEEE Microwave Theory Tech. Symp.*, V. 2 1989
45. Smith, *IEEE Microwave Theory Tech. Symp.*, V. 3 1989
46. Lee, *IEEE Microwave Theory Tech. Symp.*, V. 3 1989
47. Dumas, *IEEE Microwave Theory Tech. Symp.*, V. 1 1989
48. Kawasaki, *IEEE Microwave Theory Tech. Symp.*, V. 1 1989
49. Yuen, *Trans. Microwave Theory Tech.*, December 1989
50. Chao, *Trans. Electron Devices*, March 1989
51. Chau, *Trans. Electron Devices*, October 1989
52. Ishikawa, *IEEE Microwave Theory Tech. Symp.*, V. 3 1989
53. Yuen, *IEEE Microwave Theory Tech. Symp.*, V. 1 1989

5.4.2 Transconductance and Gate-Source Capacitance

Figures 5.38 and 5.39 present reported transconductance and gate-source capacitance as a function of gate length for HEMT devices reported in the technical literature. For reference, the least-squares fit line to the corresponding MESFET data of Figures 5.26 and 5.27 are also illustrated on the plots.

Almost all of the HEMT transconductance data reported in the technical literature lie above the least-squares fit line determined from examining MESFET data. As expected, the reported HEMT transconductance increases with decreasing gate length. Transconductance values for 0.1-μm-gate-length HEMTs have already been reported with values as high as 1160 mS/mm. High transconductance is a key characteristic of these devices, which stimulates interest in the HEMT from device and circuit designers.

Figure 5.39 presents HEMT gate-source capacitance data as a function of device gate length. The least-squares line plotted on Figure 5.39 is determined from the MESFET data of Figure 5.27 and is presented for comparison. The HEMT gate-source capacitance data are seen to be very comparable to MESFET capacitance data. Little gate length dependence is observed, and many of the data are well approximated using a 1 pF/mm constant capacitance. As in the case of the

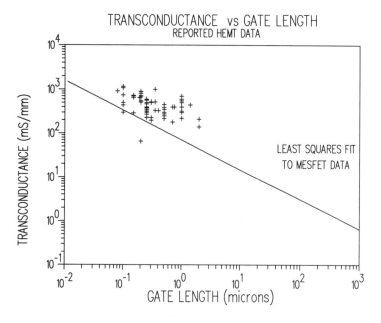

Figure 5.38 Reported transconductance per unit gate width as a function of reported HEMT gate length. The devices used to produce the figure are from the reports of Table 5.3.

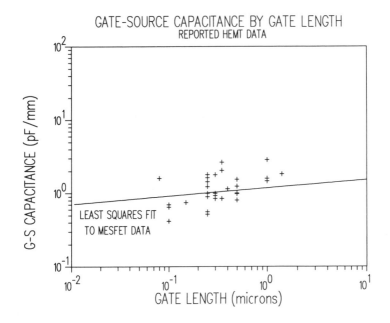

Figure 5.39 Reported gate-source capacitance per unit gate width as a function of reported HEMT gate length. The devices used to produce the figure are from the reports of Table 5.3.

MESFET, this nearly constant capacitance per unit gate width is the result of the combination of reduced gate length (acting to reduce capacitance) along with increased doping densities (acting to increase capacitance).

5.4.3   Maximum Frequency of Oscillation and Gain-Bandwidth Product

Figure 5.40 presents maximum frequency of oscillation f_{max} as a function of HEMT gate length. For comparison, the least-squares fit line to the corresponding MESFET data of Figure 5.28 is also presented on the plot. As discussed in Section 5.2, values for f_{max} can be obtained either by extrapolating measured power gain data or by computing the maximum frequency of oscillation from available equivalent circuit element values using equation (5.7) or (5.8). Data presented in Figure 5.40 are obtained by using a combination of these techniques.

The HEMT f_{max} data reported to date are seen not to exceed MESFET data dramatically—especially for short-gate-length devices. This is not anticipated from first-order device theory and may be an indication of the relative immaturity of HEMT fabrication procedures as opposed to a physical limitation of the structure.

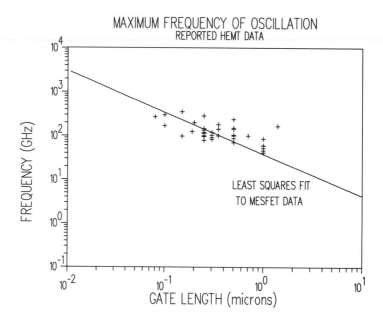

Figure 5.40 Reported or calculated maximum frequency of oscillation f_{max}, as a function of reported HEMT gate length.

Slightly higher f_{max} values are expected from HEMTs over GaAs MESFETs with equivalent gate length dimensions.

The HEMT gain-bandwidth product f_T, as a function of gate length, is p⸱ sented in Figure 5.41. As in the case of maximum frequency of oscillation, value has been obtained either by direct measurement of current gain as a funct⸱ of frequency or from equivalent circuit calculations. Most of the reported HEM⸱ data is seen to lie above the least-squares fit line determined from the MESFET data of Figure 5.29. Gain-bandwidth products as high as 210 GHz have already been reported for HEMTs with a 0.1-μm gate length. This value is approximately 5% greater than the projected ultimate achievable f_T value for the MESFET discussed in Section 5.3.

5.4.4 Noise Figure Limits

Figure 5.42 presents HEMT minimum noise figure data obtained from the reports listed in Table 5.3. As in the case of the MESFET, HEMT noise figure data are more difficult to analyze and interpret. The data make clear, however, the fact that HEMT minimum noise figures are significantly lower than those obtained in MESFETs with equivalent geometries. This is especially true at frequencies above

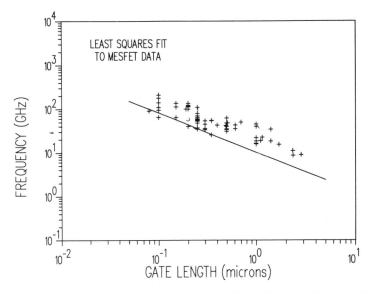

Figure 5.41 Reported or calculated gain-bandwidth product f_T, as a function of reported HEMT gate length.

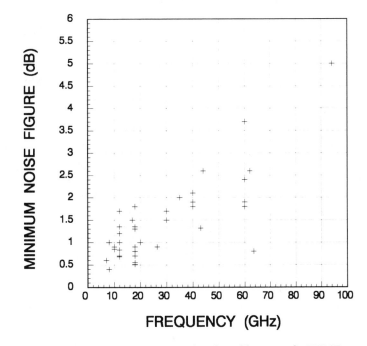

Figure 5.42 Reported minimum noise figure as a function of frequency for HEMTs.

about 20 to 30 GHz, for which very few MESFET data are available. The minimum noise figure obtained with HEMTs already approaches or surpasses the ultimate minimum noise figure projected for optimally scaled 0.1-μm MESFETs (see Figure 5.32). Figure 5.42 indicates that minimum noise figure values below 2 dB have been observed to frequencies above 60 GHz. At frequencies below 20 GHz, HEMT noise figure values approach the measurement accuracy of modern noise figure measurement equipment.

5.4.5   Output Power Limits

As in the case of the MESFET, power HEMT scaling rules are not generally compatible with high frequency, low noise scaling rules. Gate-drain avalanche breakdown is of critical concern in the design of microwave power HEMTs, as it is for power MESFETs. The high doping concentrations typically required for high frequency HEMT fabrication act to reduce device breakdown voltage and limit HEMT power performance characteristics. This constraint favors lower doping levels for power devices than for the devices utilized in high frequency or low noise applications. Unlike the MESFET, however, obtaining sufficient carrier densities in the electron gas channel can also be a critical concern for HEMT power devices. Sufficient current must flow in the device channel to support the desired RF power levels. This leads to requirements for higher doping densities, in direct conflict with the breakdown voltage constraint. In practice, this has led device designers to more sophisticated HEMT structures.

For the reasons mentioned in the previous paragraph, HEMTs fabricated using a conventional single AlGaAs/GaAs layer structure tend to exhibit inferior power characteristics [49]. Multiple quantum-well and pseudomorphic structures are much better suited for applications involving significant power densities. In the multiple quantum-well device, increased sheet carrier density is achieved by utilizing more than one parallel active channel. This allows higher current carrying capability to be achieved while maintaining lower doping concentrations in the doped AlGaAs layer. The lower doping levels contribute to higher breakdown voltage and, therefore, higher maximum output powers. In pseudomorphic structures, increased carrier concentrations are obtained because the dimensions of the quantum well are increased over those in the conventional HEMT. The depth (in terms of electron energy) of the well is increased by utilizing narrower band-gap material than GaAs, while the width of the well is increased slightly by control of the small-band-gap material layer thickness. Again, these improvements provide lower doping density requirements and associated higher breakdown voltages. The InP-based pseudomorphic HEMT structure offers the additional advantage of providing higher thermal conductivity than GaAs. This maintains a lower junction temperature in the device, which favors high power applications.

All of the power HEMT data collected and presented in this section are taken from either multiple quantum-well or pseudomorphic structures. Table 5.4 lists the references used to compile the power HEMT data that follow.

Table 5.4
Recent Published Maximum Output Power Reports for HEMTs*

1. Saunier, *IEEE Electron Devices Lett.*, September 1986
2. Sovero, *IEEE Trans. Electron Devices*, October 1986
3. Hikosaka, *IEEE Electron Devices Lett.*, November 1987
4. Saunier, *IEEE Electron Devices Lett.*, August 1988
5. Smith, *Appl. Microwave*, May 1989
6. Kao, *IEEE Trans. Electron Devices*, December 1989
7. Ferguson, *IEEE Trans. Microwave Theory Tech. Symp.*, 1989

*Reported data are for Figures 5.43 through 5.48.

Definition of a single FOM to summarize the potential of a HEMT for power applications is not possible. As with the MESFET, increases in output power are associated with decreases in both gain and efficiency when the device is operating near maximum output power conditions. These various performance characteristics must be balanced according to the requirements for a particular application.

Figure 5.43 presents maximum output power per unit gate width as a function of frequency for the HEMT devices described in the reports listed in Table 5.4. The devices are divided into three different gate length categories. The tendency for short-gate-length devices to exhibit higher power densities at higher frequencies is evident. For comparison, the 0.5-μm-gate-length HEMT data are plotted along with MESFET data for corresponding gate length devices in Figure 5.44. The MESFET data are identical to those plotted in Figure 5.35. The data of Figure 5.44 indicate that HEMT devices produce output power densities comparable to MESFETs.

Reported power-added efficiency is presented in Figure 5.45 for the HEMT devices of Table 5.4. The trend for shorter gate length devices to exhibit superior performance to longer gate length devices is clearly illustrated. Likewise, efficiency is seen to decrease with increasing frequencies for all gate length categories. The 0.5-μm power HEMT efficiency data are plotted again in Figure 5.46 along with the corresponding MESFET data. The HEMT power-added efficiencies are seen to be very comparable to the MESFET efficiency data.

Figure 5.47 presents reported power gain under maximum output power conditions for the three gate length categories of HEMTs. Again, the expected trends with frequency and gate length are observed. For comparison, Figure 5.48 presents

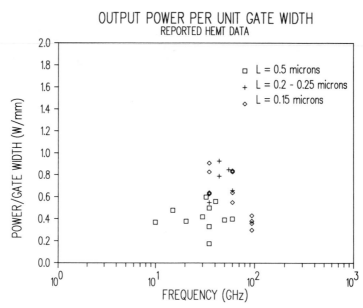

Figure 5.43 Reported output power per unit gate width as a function of frequency for the HEMTs listed in Table 5.4.

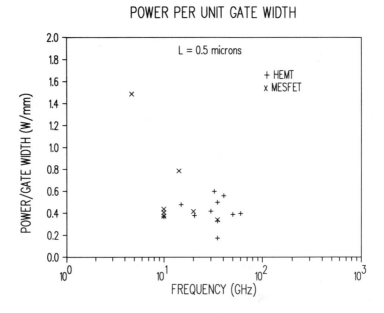

Figure 5.44 Reported output power per unit gate width as a function of frequency for the 0.5-μm-gate-length MESFETs and HEMTs listed in Tables 5.2 and 5.4.

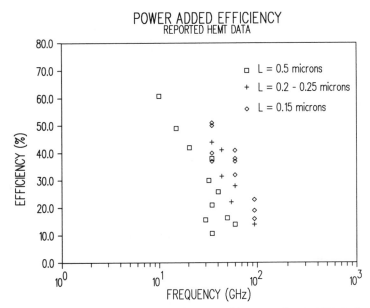

Figure 5.45 Reported power-added efficiency as a function of frequency for the HEMTs listed in Table 5.4.

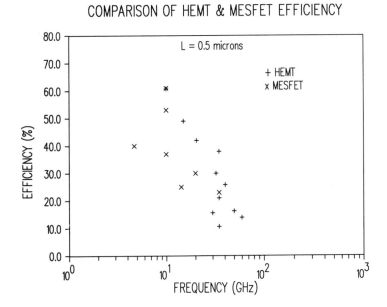

Figure 5.46 Reported power-added efficiency as a function of frequency for 0.5-μm-gate-length MES-FETs and HEMTs listed in Tables 5.2 and 5.4.

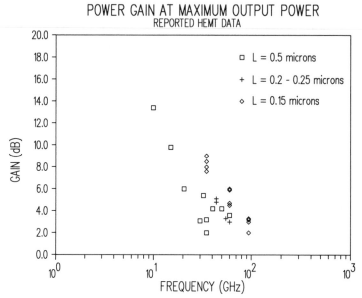

Figure 5.47 Reported power gain as a function of frequency for the HEMTs listed in Table 5.4.

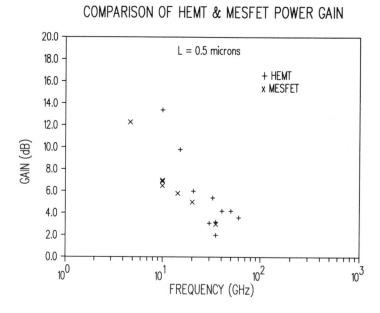

Figure 5.48 Reported power gain as a function of frequency for the 0.5-μm MESFETs and HEMTs listed in Tables 5.2 and 5.4.

gain as a function of frequency for both 0.5-μm MESFETs and HEMTs. The best HEMT data are seen to be superior to the best MESFET data at all frequencies.

The data illustrated in Figures 5.43 through 5.48 indicate that power density and power-added efficiency values for HEMTs are expected to be comparable to those obtained using MESFETs (i.e., maximum output power densities greater than about 1 W/mm and power-added efficiencies in the 40 to 60% range). Associated gains for the HEMTs operating under these conditions will be higher by 2 to 3 dB than for the MESFET in similar operating circumstances. As with the MESFET, greater efficiency or gain values can be achieved at the expense of total output power.

REFERENCES

[1] J.J. Komiak, "Wideband HEMT Balanced Amplifier," *Electron. Lett.,* Vol. 22, No. 14, July 1986, pp. 747–749.

[2] J. Eisenberg, W. Ou, and J. Archer, "A 0.5-6 GHz High Gain Low Noise MMIC Amplifier," *IEEE GaAs IC Symp. Digest,* 1989, pp. 83–86.

[3] K.B. Niclas and R.R. Pereira, "Performance of a 2–18 GHz Ultra Low-Noise Amplifier Module," *IEEE Microwave Theory Tech. Symp. Digest,* 1989, pp. 841–844.

[4] S. Bandla, G. Dawe, C. Bedard, R. Tayrani, D. Shaw, L. Raffaelli, and R. Goldwasser, "A 35 GHz Monolithic MESFET LNA," *IEEE Microwave Millimeter-Wave Monolithic Circuits Symp. Digest,* 1988, pp. 151–155.

[5] H.L.A. Hung, T.T. Lee, F.R. Phelleps, J.F. Singer, J.F. Bass, T.F. Noble, and H.C. Huang, "60 GHz GaAs MMIC Low Noise Amplifiers," *IEEE Microwave Millimeter-Wave Monolithic Circuits Symp. Digest,* 1988, pp. 87–90.

[6] K.H.G. Duh, P.C. Chao, P. Ho, A. Tessmer, S.M.J. Liu, M.Y. Kao, P.M. Smith, and J.M. Ballingal, "W-Band InGaAs HEMT Low Noise Amplifiers," *IEEE Microwave Theory Tech. Symp. Digest,* 1990, pp. 595–598.

[7] K.H. Duh, P.C. Chao, P.M. Smith, L.F. Lester, B.R. Lee, L.M. Ballingal, and M-Y. Kao, "High Performance Ka-Band and V-Band HEMT Low Noise Amplifiers," *IEEE Trans. Microwave Theory Tech.,* Vol. MTT-365, No. 12, December 1988, pp. 1598–1603.

[8] C.L. Lau, M. Feng, G.W. Wang, T. Lepkowski, Y. Chang, C. Ito, V. Dunn, N. Hodges, and J. Schellenberg, "44 GHz Hybrid Low Noise Amplifiers Using Ion-Implanted In$_x$-Ga$_{1-x}$As MESFETs," *IEEE Microwave Theory Tech. Symp. Digest,* 1990, pp. 431–433.

[9] I.J. Bahl, E.L. Griffen, A.E. Geissberger, C. Andricos, and T.F. Brukiewa, "Class-B Power MMIC Amplifiers with 70 Percent Power-Added Efficiency," *IEEE Trans. Microwave Theory Tech.,* Vol. MTT-37, No. 9, September 1989, pp. 1315–1320.

[10] S. Arai, T. Yoshida, K. Kai, S. Takatsuka, Y. Oda, and S. Yanagawa, "Ka-Band 2 Watt Power GaAs MMIC," *IEEE Microwave Theory Tech. Symp. Digest,* 1989, pp. 1105–1108.

[11] D.W. Ferguson, P.M. Smith, P.C. Chao, L.F. Lester, R.P. Smith, P. Ho, A. Jabra, and J.M. Ballingal, "44 GHz Hybrid HEMT Power Amplifiers," *IEEE Microwave Theory Tech. Symp. Digest,* 1989, pp. 987–990.

[12] R.D. Boesch and J.A. Thompson, "X-Band 0.5, 1, and 2 Watt Power Amplifiers with Marked Improvement in Power Added Efficiency," *IEEE Trans. Microwave Theory Tech.,* Vol. MTT-38, No. 6, June 1990, pp. 707–711.

[13] S. Toyoda, "Broadband Push-Pull Power Amplifier," *IEEE Microwave Theory Tech. Symp. Digest,* 1990, pp. 507–510.

[14]   M.A. Khatibzadeb and H.Q. Tserng, "Harmonic Tuning of Power FETs at X-Band," *IEEE Microwave Theory Tech. Symp. Digest*, 1990, pp. 989–992.

[15]   Y. Taniguchi *et al.*, "A C-Band 25 Watt Linear Power Amplifier," *IEEE Microwave Theory Tech. Symp. Digest*, 1990, pp. 981–984.

[16]   R.S. Pengelly, *Microwave Field-Effect Transistors—Theory, Design and Applications*, Chichester, UK: Research Studies Press, 1982, pp. 245–298.

[17]   J.M. Golio and C.M. Krowne, "New Approach for FET Oscillator Design," *Microwave J.*, October 1978, pp. 59–61.

[18]   P. Bura and R. Dikshit, "FET Mixers For Communication Satellite Transponders," *IEEE Microwave Theory Tech. Symp. Digest*, 1976, pp. 90–92.

[19]   P. Bura and B. Vassilakis, "A Balanced 11 GHz HEMT Up-Converter," *IEEE Microwave Theory Tech. Symp. Digest*, Vol. III, 1989, pp. 1299–1302.

[20]   V. Hwang and T. Itoh, "Quasi-Optical HEMT and MESFET Self-Oscillating Mixers," *IEEE Trans. Microwave Theory Tech.*, Vol. MTT-36, December 1988, pp. 1701–1705.

[21]   P.D. Chow, D. Garske, J. Velebir, E. Hsieh, Y.C. Nhan, and H.C. Yen, "Design and Performance of a 94 GHz HEMT Mixer," *IEEE Microwave Theory Tech. Symp. Digest*, Vol. II, 1989, pp. 731–734.

[22]   S. Weiner, D. Neuf, and S. Spohrer, "2 To 8 GHz Double Balanced MESFET Mixer with +30 dBm 3rd Order Intercept," *IEEE Microwave Theory Tech. Symp. Digest*, Vol. II, 1988, pp. 1097–1100.

[23]   J.B. Tsui, *Microwave Receivers and Related Components*, Los Altos, CA: Peninsula Publishing, 1985.

[24]   H. Kondoh, "DC-50 GHz Variable Attenuator with a 30 dB Dynamic Range," *IEEE Microwave Theory Tech. Symp. Digest*, Vol. I, 1988, pp. 499–502.

[25]   J. Bayruns, P. Wallace, and N. Scheinberg, "A Monolithic DC-1.6 GHz Digital Attenuator," *IEEE Microwave Theory Tech. Symp. Digest*, Vol. III, 1989, pp. 1295–1298.

[26]   G. Lizama, T. Andrade, and R. Benton, "1–6 GHz GaAs MMIC Linear Attenuator with Integral Drivers," *IEEE Microwave Millimeter-Wave Monolithic Circuits Symp.*, 1987, pp. 105–107.

[27]   Y. Tajima, T. Tsukii, R. Mozzi, E. Tong, L. Hanes, and B. Wrona, "GaAs Monolithic Wideband (2–18 GHz) Variable Attenuators," *IEEE Microwave Theory Tech. Symp. Digest*, 1982, pp. 479–481.

[28]   P.L. Hower, W.W. Hooper, B.R. Cairns, R.D. Fairman, and D.A. Treme, "The GaAs Field Effect Transistor," in *Semiconductors and Semimetals*, Vol. 7A, ed. Willardson and Beer, New York: Academic Press, 1971, pp. 147–200.

[29]   R.J. Trew, M.B. Steer, and D. Chamberlain, "Parasitic Effects Upon the High-Frequency Performance of Three-Terminal Devices," presented at 3rd GaAs Simulation Workshop, University of Duisburg, October 7–8, 1986.

[30]   P.H. Ladbrooke, *MMIC Design: GaAs FETs and HEMTs*, Norwood, MA: Artech House, 1989, pp. 223–258.

[31]   R.J. Trew, *Microwave Solid State Circuit Design*, Chap. 7, ed. I.J. Bahl and P. Bharita, New York: John Wiley and Sons, 1988.

[32]   C.A. Liechti, "Microwave Field-Effect Transistors—1976," *IEEE Trans. Microwave Theory Tech.*, Vol. MTT-24, June 1976, pp. 279–299.

[33]   J.V. DiLorenzo and D.D. Khandelwal, *GaAs FET Principles and Technology*, Norwood, MA: Artech House, 1982, pp. 431–439.

[34]   J.M. Golio, "Ultimate Scaling Limits for High-Frequency GaAs MESFETs," *IEEE Trans. Electron Devices*, Vol. ED-35, July 1988, pp. 839–848.

[35]   K. Yoloyama, M. Tomizawa, and A. Yoshi, "Scaled Performance for Submicron GaAs MESFET's," *IEEE Electron Devices Lett.*, Vol. EDL-6, October 1985, pp. 536–538.

[36]   M.F. Abusaid and J.R. Hauser, "Calculations of High-Speed Performance for Submicrometer

Ion-Implanted GaAs MESFET Devices," *IEEE Trans. Electron Devices,* Vol. ED-33, July 1986, pp. 913–918.

[37] H. Daemkes, W. Brocherhoff, K. Heime, and A. Cappy, "Improved Short-Channel GaAs MES-FET's by Use of Higher Doping Concentrations," *IEEE Trans. Electron Devices,* Vol. ED-31, August 1984, pp. 1032–1037.

[38] M. Reiser and P. Wolf, "Computer Study of Submicrometer FET's," *Electron. Lett.,* Vol. 8, May 18, 1972, pp. 254–256.

[39] K.E. Drangeid and R. Sommerhalder, "Dynamic Performance of Schottky-barrier Field-effect Transistors," *IBM J. Res. Dev.,* Vol. 14, March 1970, pp. 82–94.

[40] W.R. Frensley, "Power-Limiting Breakdown Effects in GaAs MESFET's," *IEEE Trans. Electron Devices,* Vol. ED-28, August 1981, pp. 962–970.

[41] M.P. Zaitlin, "Reverse Breakdown in GaAs MESFET's," *IEEE Trans. Electron Devices,* Vol. ED-33, November 1986, pp. 1635–1639.

[42] S.H. Wemple, W.C. Niehaus, H.M. Cox, J.V. DiLorenzo, and W.O. Schlosser, "Control of Gate-Drain Avalanche in GaAs MESFET's," *IEEE Trans. Electron Devices,* Vol. ED-27, June 1980, pp. 1013–1018.

[43] P. Canfield, J. Medinger, and L. Forbes, "Buried-Channel GaAs MESFET's with Frequency-Independent Output Conductance," *IEEE Electron Devices Lett.,* Vol. EDL-8, March 1987, pp. 88–89.

[44] H. Fukui, "Design of Microwave GaAs MESFETs for Broadband Low-Noise Amplifiers," *IEEE Trans. Microwave Theory Tech.,* Vol. MTT-27, July 1979, pp. 643–650.

[45] B. Kim, N. Camilleri, H.D. Shih, H.Q. Tserng, and M. Wurtele, "35 GHz GaAs Power MESFETs and Monolithic Amplifiers," *IEEE Trans. Microwave Theory Tech.,* Vol. MTT-37, September 1989, pp. 1327–1333.

[46] D. Bechtle, J. Klatskin, G. Taylor, M. Eron, S.G. Liu, R.L. Camisa, and H. Dudley, "K and Ka-Band High Efficiency Amplifier Modules Using GaAs Power FETs," *IEEE Microwave Theory Tech. Symp. Digest,* 1987, pp. 849–851.

[47] M.R. Weiss and D. Pavlidis, "An Investigation of the Power Characteristics and Saturation Mechanisms in HEMT's and MESFET's," *IEEE Trans. Electron Devices,* Vol. ED-35, August 1988, pp. 1197–1206.

[48] J.R. Hauser, "Avalanche Breakdown Voltages for III-V Semiconductors," *Appl. Phys. Lett.,* Vol. 33, August 1978, pp. 351–354.

[49] P.M. Smith and A.W. Swanson, "HEMTs—Low Noise and Power Transistors for 1 to 100 GHz," *Appl. Microwave,* Vol. 1, May 1989, pp. 63–72.

LIST OF SYMBOLS

$A_0, \ldots,$ A_3	Empirical transconductance parameters used in the Curtice-Ettenberg model
a	Epitaxial layer thickness of a MESFET
B	Bandwidth of a device, circuit or system
b	Empirical transconductance modification term used in the Statz model
C_{ds}	Drain-to-source capacitance of a device
C_{gd}	Gate-to-drain capacitance of a device
C_{GDI}	Parameter used in the semiempirical capacitance model
C_{GD0}	Empirical parameter used in HEMT and semiempirical capacitance models
C_{gd0}	Empirical capacitor model parameter representing the zero bias gate-to-drain capacitance of a device
C_{gs}	Gate-to-source capacitance of a device
C_{GS0}	Empirical parameter used in HEMT and semiempirical capacitance models
C_{gs0}	Empirical capacitor model parameter representing the zero bias gate-to-source capacitance of a device
C_{m0}	Empirical total gate electrode capacitance parameter for HEMT capacitance model
D_n	Electron diffusion coefficient in n-type semiconductor
d	Layer thickness of the doped wide-band-gap semiconductor material used to fabricate a HEMT
d_0	Layer thickness of the undoped spacer layer used to fabricate a HEMT
E	Electric field intensity
E_c	Energy level of the conduction band in a semiconductor
E_F	Fermi energy level in a semiconductor
E_G	Maximum gain efficiency of a device
E_g	Energy gap of a semiconductor
EMU	Error magnitude unit used as an objective function for large-signal model parameter extraction optimizations
E_v	Energy level of the valence band in a semiconductor

F_{dB}	Noise figure expressed in dB
F_{min}	Minimum noise figure for a device
f_{max}	Maximum frequency of oscillation of a device
f_T	Gain-bandwidth product of a device
G	Gain of a two port network (or device)
G_{ds}	The dc output conductance of device
G_m	The dc transconductance of a device
g_{ds}	RF or modeled output conductance of device
g_m	RF or modeled transconductance of a device
I_{ds}	Drain-to-source current of a device
I_{dss}	Saturated drain-to-source current of a device with 0 volts gate-to-source bias applied
I_g	Current flowing through the gate terminal of a MESFET or HEMT
I_s	Reverse saturation current for a rectifying contact on a semiconductor sample
i_d	Equivalent noise current at the drain of the MESFET or HEMT
i_g	Noise current induced in the gate circuit by charge fluctuations in the drain current
J_{diff}	Current density due to electron diffusion in a semiconductor sample
J_{drift}	Current density due to electron drift in a semiconductor sample
J_n	Total current density due to electrons in a semiconductor sample
k	Boltzmann's constant
k_f	An empirical fitting factor for the Cappy noise model
$k_1, \ldots,$ k_4	Empirical fitting factors for the Fukui noise model
L	Gate length of a MESFET or HEMT
L_D	Debye length
L_d	Parasitic drain inductance or when used as a dimension, the length of the drain contact
L_g	Parasitic gate inductance
L_s	Parasitic source inductance or when used as a dimension, the length of the source contact
m_{GD}	Parameter used in semiempirical capacitance model
m_{GS}	Parameter used in semiempirical capacitance model
MSG	Maximum stable gain of a device
N_a	Concentration of background acceptor levels (p-type doping density)
N_c	Density of states in the conduction band of a semiconductor
N_d	Concentration of background donor levels (n-type doping density)
N_i	Noise power at the input port of a two-port network
N_o	Noise power at the output port of a two-port network
N_T	Concentration of traps (deep level density)
N_v	Density of states in the valence band of a semiconductor

n	Concentration of free electrons in a semiconductor
n_s	Sheet carrier density of the 2DEG in a HEMT
n_{so}	Maximum possible sheet carrier density for a HEMT heterostructure
P_N	Maximum noise power generated by a resistor
p	Concentration of holes in a semiconductor
q	Electronic charge
R_d	Parasitic drain resistance of a device
R_{ds}	The dc output resistance of a device
R_g	Parasitic gate resistance of a device
R_i	Equivalent charging resistance of device
R_n	Equivalent noise resistance of a device
R_s	Parasitic source resistance of a device
r_{ds}	RF or modeled RF output resistance of a device
S_i	Signal power at the input port of a two-port network
S_{io}	Noise power spectral density of a device
S_o	Signal power at the output port of a two-port network
T_N	Noise temperature of a device or network
T_0	Room temperature (290 K)
V_{bi}	Built-in potential of a rectifying contact
V_{br}	Breakdown voltage of a semiconductor sample or device
V_{ds}	Drain-to-source voltage
V_{DSAT}	Drain-to-source saturation voltage
V_{dso}	Drain-to-source voltage (in saturation) at which the A_i coefficients are evaluated in the Curtice-Ettenberg model
V_{GC}	Critical gate potential at which the electron concentration in the 2-DEG channel reaches the maximum possible sheet density
V_{gs}	Gate-to-source voltage
V_N	Root mean, square (rms) thermal noise voltage of a resistor
V_{pf}	Empirical HEMT model parameter specifying the gate voltage where transconductance begins to degrade
V_T, V_{T0}	Pinch-off or threshold voltage of a MESFET or HEMT
v	Carrier drift velocity
Y_{opt}	Optimum admittance required at the input of a device to achieve minimum noise figure ($Y_{opt} = G_{opt} + jB_{opt}$)
Z	Gate width of a MESFET or HEMT
α	Empirical noise current factor used in the Cappy noise model and empirical current saturation factor used in a number of large-signal models
β	Empirical noise current factor used in the Cappy noise model and empirical transconductance factor used in a number of large-signal models
β_{GS}	Parameter used in semiempirical capacitance model
Γ	Voltage reflection coefficient
Γ_{opt}	Input reflection coefficient required to achieve minimum noise figure

δ	Noise temperature *versus* electric field curve fitting factor, used in the Pucel noise model, as an empirical capacitance parameter in the Statz model, and as an empirical transconductance parameter in the TOM model
ϵ	Dielectric constant of a material
γ	Empirical output conductance factor used in the TOM model
λ	Empirical output conductance factor used in several large-signal models
λ_{GS1}	Parameter used in semiempirical capacitance model
λ_{GS2}	Parameter used in semiempirical capacitance model
λ_{GD1}	Parameter used in semiempirical capacitance model
λ_{GD2}	Parameter used in semiempirical capacitance model
μ_{crit}	Empirical parameter representing the critical field for mobility degradation in the advanced Curtice model
μ_n	Low-field mobility of electrons in n-type semiconductor
μ_0	Low-field mobility of carriers in an undepeleted MESFET channel (equivalent to μ_n for typical n-channel devices)
ν	Empirical velocity saturation parameter for the Lehovec and Zuleeg model
ξ	Empirical parameter used for HEMT models
τ	Transit time delay associated with carriers traveling across a MESFET or HEMT channel
υ_{sat}	Saturated drift velocity parameter used in the Lehovec and Zuleeg model
χ	Empirical gate electrode coefficient parameter used in HEMT capacitance models
ψ	Empirical parameter used for HEMT models
σ	Semiconductor conductivity

INDEX

THE AUTHORS

Eric N. Arnold was born in Chicago, Illinois, in 1962. He received his BS and MS degrees in electrical engineering from Arizona State University, Tempe, in 1985 and 1988, respectively. As a research assistant, Mr. Arnold worked in the area of nonlinear device circuit modeling. In 1988, Mr. Arnold joined Motorola's Government Electronics Group in Chandler, Arizona, where he worked on MMIC amplifier design and the development of an in-house parameter extraction tool for GaAs FETs. Early in 1990, Mr. Arnold joined EEsof, where he is involved in the development of accurate nonlinear models for use in harmonic balance and SPICE-based simulators. In the course of his work, Mr. Arnold has authored or co-authored several technical papers related to time-domain circuit modeling and model parameter extraction.

William B. Beckwith received his BEE degree from the Georgia Institute of Technology in 1984, and is currently completing requirements for an MSEE degree at Arizona State University. Since 1984, Mr. Beckwith has been with the Motorola Government Electronics Group. His responsibilities have included the design of broadband power dividers and couplers as well as various MMIC circuits. Mr. Beckwith has also been heavily involved in the development of passive device models for use in MMIC circuit design, and holds two patents for unique circuits implemented by using MMIC technology.

J. Michael Golio received his BSEE degree from the University of Illinois in 1976 and MSEE and PhD degrees from North Carolina State University in 1980 and 1983, respectively. From 1976 to 1978, Dr. Golio worked with Watkins-Johnson Company in the Microwave Tunable Devices Group. His responsibilities included the design of microwave oscillators and automated measurement systems. Dr. Golio's graduate work at North Carolina State University was in the area of microwave devices and semiconductor transport. After two years as a faculty member at Arizona State University, Dr. Golio joined Motorola's Government Electronics Group. His primary responsibility there was the direction of a research effort on large signal microwave circuit design. Dr. Golio was also involved in research on high-temperature superconductors and optical signal distribution. In 1990, Dr. Golio joined Motorola's Semiconductor Research and Development Laboratory as Characterization Manager.

Monte Miller received his BS degree in electrical engineering technology from Arizona State University, Tempe, in 1985, and is currently working toward his MSEE degree also at Arizona State. Mr. Miller joined Motorola Government Electronics Group, Chandler, Arizona, in 1985, where his responsibilities involve device characterization and modeling and circuit simulation.

Joseph Staudinger received a BSEE degree from Kansas State University and his MSEE degree from Arizona State University, Tempe, in 1980 and 1987, respectively. Since 1981, Mr. Staudinger has been employed by the Government Electronics Group of Motorola, working in the area of design and development of MIC and MMIC components, integrated subsystems, and related packaging issues. Since 1987, Mr. Staudinger has been primarily responsible for the direction of research efforts in developing highly integrated functions contained on single MMIC die. He has developed numerous MMIC components including single-chip down-converters, amplifiers, balanced mixers, multipliers, switches, power dividers, and numerous other devices. Mr. Staudinger is a member of IEEE, has authored several manuscripts on MIC and MMIC component design techniques, and has several patents pending.

The Artech House Microwave Library

LOSLIN: Lossy Line Calculation Software and User's Manual, Fred E. Gardiol

Lossy Transmission Lines, Fred E. Gardiol

MATCHNET: Microwave Matching Networks Synthesis, Stephen V. Sussman-Fort

Materials Handbook for Hybrid Microelectronics, J.A. King, ed.

Matrix Parameters for Multiconductor Transmission Lines: Software and User's Manual, A.R. Djordjevic, et al.

MIC and MMIC Amplifier and Oscillator Circuit Design, Allen Sweet

Microelectronic Reliability, Volume I: Reliability, Test, and Diagnostics, Edward B. Hakim, ed.

Microelectronic Reliability, Volume II: Integrity Assessment and Assurance, Emiliano Pollino, ed.

Microstrip Antenna Design, K.C. Gupta and A. Benalla, eds.

Microstrip Lines and Slotlines, K.C. Gupta, R. Garg, and I.J. Bahl

Microwave Circulator Design, Douglas K. Linkhart

Microwave Engineers' Handbook: 2 volume set, Theodore Saad, ed.

Microwave Filters, Impedance Matching Networks, and Coupling Structures, G.L. Matthaei, L. Young, and E.M.T. Jones

Microwave Integrated Circuits, Jeffrey Frey and Kul Bhasin, eds.

Microwaves Made Simple: Principles and Applications, Stephen W. Cheung, Frederick H. Levien, *et al.*

Microwave Materials and Fabrication Techniques, Second Edition, Thomas S. Laverghetta

Microwave MESFETs and HEMTs, J. Michael Golio, et al.

Microwave and Millimeter Wave Heterostructure Transistors and Applications, F. Ali, ed.

Microwave Mixers, Stephen A. Maas

Microwave Transmission Design Data, Theodore Moreno

Microwave Transition Design, Jamal S. Izadian and Shahin M. Izadian

Microwave Transmission Line Filters, J.A.G. Malherbe

Microwave Transmission Line Couplers, J.A.G. Malherbe

Microwave Tubes, A.S. Gilmour, Jr.

MMIC Design: GaAs FETs and HEMTs, Peter H. Ladbrooke

Modern GaAs Processing Techniques, Ralph Williams

Modern Microwave Measurements and Techniques, Thomas S. Laverghetta

Monolithic Microwave Integrated Circuits: Technology and Design, Ravender Goyal, *et al.*

Nonlinear Microwave Circuits, Stephen A. Maas

Optical Control of Microwave Devices, Rainee N. Simons

PLL: Linear Phase-Locked Loop Control System Analysis Software and User's Manual, Eric L. Unruh

Receiving Systems Design, Stephen J. Erst

Scattering Parameters of Microwave Networks with Multiconductor Transmission Lines: Software and User's Manual, A.R. Djordjevic, et al.

Solid-State Microwave Devices, Thomas S. Laverghetta

Stripline Circuit Design, Harlan Howe, Jr.

Terrestrial Digital Microwave Communications, Ferdo Ivanek, *et al.*

Time-Domain Response of Multiconductor Transmission Lines: Software and User's Manual, A.R. Djordjevic, et al.